Linux for Embedded and Real-Time Applications

Linux for Embedded and Real-Time Applications

Third Edition

Doug Abbott

AMSTERDAM • BOSTON • HEIDELBERG • LONDON • NEW YORK • OXFORD
PARIS • SAN DIEGO • SAN FRANCISCO • SINGAPORE • SYDNEY • TOKYO
Newnes is an imprint of Elsevier

Newnes is an imprint of Elsevier
225 Wyman Street, Waltham, 02451, USA
The Boulevard, Langford Lane, Kidlington, Oxford OX5 1GB, UK

First edition 2003
Second edition 2006
Third edition 2013

Notice
Knowledge and best practice in this field are constantly changing. As new research and experience broaden our understanding, changes in research methods, professional practices, or medical treatment may become necessary.

Practitioners and researchers must always rely on their own experience and knowledge in evaluating and using any information, methods, compounds, or experiments described herein. In using such information or methods they should be mindful of their own safety and the safety of others, including parties for whom they have a professional responsibility.

To the fullest extent of the law, neither the Publisher nor the authors, contributors, or editors, assume any liability for any injury and/or damage to persons or property as a matter of products liability, negligence or otherwise, or from any use operation of any methods, products, instructions, or ideas contained in the material herein.

Library of Congress Cataloging-in-Publication Data
A catalog record for this book is available from the Library of Congress

British Library Cataloguing-in-Publication Data
A catalogue record for this book is available from the British Library

ISBN: 978-0-12-415996-9

For information on all Newnes publications
visit our web site at www.newnespress.com

Printed and bound in Great Britain
12 11 10 9 8 7 6 5 4 3 2 1

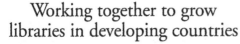

Working together to grow
libraries in developing countries

www.elsevier.com | www.bookaid.org | www.sabre.org

ELSEVIER   BOOK AID International   Sabre Foundation

Dedication

On a personal level, this book is dedicated to the two most important people in my life:

To Susan, my best friend, my soul mate. Thanks for sharing life's journey with me.
To Brian, budding lighting designer, actor and musician, future pilot, and all-around great kid. Thanks for keeping me young at heart.

Never doubt that a small band of committed citizens can change the world. Indeed, it is the only thing that ever has.

Margaret Mead

On a professional level, this book is dedicated to those who made it possible; Open Source programmers everywhere, especially that small band of Open Source pioneers whose radical notion of free software for the good of the community, and just for the fun of it, has indeed changed the world.

Contents

Preface

It was twenty years ago today
Sergeant Pepper taught the band to play
John Lennon, Sergeant Pepper's Lonely Hearts Club Band

```
From: torvalds@klaava.Helsinki.FI (Linus Benedict Torvalds)

Newsgroups: comp.os.minix

Subject: What would you like to see most in minix?

Summary: small poll for my new operating system

Message-ID: <1991Aug25.205708.9541@klaava.Helsinki.FI>

Date: 25 Aug 91 20:57:08 GMT

Organization: University of Helsinki

Hello everybody out there using minix -

I'm doing a (free) operating system (just a hobby, won't be big and professional like gnu)
for 386(486) AT clones. This has been brewing since April, and is starting to get ready. I'd
like any feedback on things people like/dislike in minix, as my OS resembles it somewhat
(same physical layout of the file-system (due to practical reasons) among other things).

I've currently ported bash(1.08) and gcc(1.40), and things seem to work. This implies that
I'll get something practical within a few months, and I'd like to know what features most
people would want. Any suggestions are welcome, but I won't promise I'll implement them:-)

Linus (torvalds@kruuna.helsinki.fi)

PS. Yes - it's free of any minix code, and it has a multi-threaded fs.

It is NOT protable (uses 386 task switching etc.), and it probably never will support
anything other than AT-harddisks, as that's all I have:-(.
```

With this message on the then-nascent, text-only Internet, Linus Torvalds, a college student
in Finland, started a revolution. As I write this in late 2011, Linux has celebrated its
20th anniversary. I doubt that Linus had the slightest clue that his "hobby" project would
soon morph into a powerful, 32-bit, and later 64-bit, operating system that credibly
competes with even the Borg of Redmond.

Needless to say, Linux has come a long way in 20 years. The original release, version 0.01 released on September 17, 1991, consisted of 88 files adding up to 278 KB. A contemporary Linux release contains close to 40,000 files taking up 400 plus MB. Literally, thousands of programmers all over the planet are engaged in Linux development in one way or another.

Much has happened in the Linux world since the second edition of this book was published in 2006, which of course is the motivation for a new edition. The kernel continues to evolve and improve having recently been rolled to version 3.0 in honor of the 20th anniversary. Google introduced the Android variant in late 2007. In August of 2011, it was estimated that Android had a 48% share of the smart phone market with over 500,000 devices being activated *every day*!

I began the preface to the first edition by confessing that I've never really liked Unix, considering it deliberately obscure and difficult to use. Initially, Linux did little to change that impression and I still have something of a love/hate relationship with the thing.

But, while Linux is still slowly getting ready for prime time in the world of desktop computing, it has achieved notable success in the embedded space where I work and in the world of network servers. In 2007, LinuxDevices.com reported Linux being used in just under 50% of embedded projects and/or products with projections of up to 70% by 2012. Even Microsoft admits that Linux runs 60% of the world's web servers vs. 40% running Windows server.

Linux is indeed complex and, unless you're already a Unix guru, the learning curve is quite steep. The information is out there on the web but it is often neither easy to find nor readable. There are probably hundreds of books in print on Linux covering every aspect from beginners' guides to the internal workings of the kernel. By now there are even a number of books discussing Linux in the embedded and real-time space.

I decided to climb the Linux learning curve partly because I saw it as an emerging market opportunity and partly because I was intrigued by the Open Source development model. The idea of programmers all over the world contributing to the development of a highly sophisticated operating system just for the fun of it is truly mind-boggling. Having the complete source code not only allows you to modify it to your heart's content but also allows you (in principle at least) to understand how the code works.

Open Source has the potential to be a major paradigm shift in how our society conducts business because it demonstrates that cooperation can be as useful in developing solutions to problems as competition. And while I doubt that anyone is getting fabulously wealthy off of Linux, a number of companies have demonstrated that it's possible to build a viable business model around it.

Audience and Prerequisites

My motivation for writing this book was to create the kind of book I wished I had when I started out with Linux. With that in mind, the book is directed at two different audiences:

1. The primary audience is embedded programmers who need an introduction to Linux in the embedded space. This is where I came from and how I got into Linux; so, it seems like a reasonable way to structure the book.
2. The other audience is Linux programmers who need an introduction to the concepts of embedded and real-time programming.

Consequently, each group will likely see some material that is review although it may be presented with a fresh perspective.

This book is not intended as a beginners' guide. I assume that you have successfully installed a Linux system and have at least played around with it some. You know how to log in, you've experimented with some of the command utilities and have probably fired up a GUI desktop. Nonetheless, Chapter 2 takes you through the installation process and Chapter 3 is a cursory introduction to some of the features and characteristics of Linux that are of interest to embedded and real-time programmers.

The book is loosely divided into three parts. Part 1 is largely introductory and sets the stage for Part 2, which discusses application programming in a cross-development environment. Part 3 takes a more detailed look at some of the components and tools introduced earlier. It also examines real-time performance in Linux.

It goes without saying that you can't learn to program by reading a book. You have to do it. That's why this book is designed as a practical hands-on guide. You'll be installing a number of software packages on your Linux workstation, and there is sample code available at the book's web site, www.elsevier.com/.

Embedded programming generally implies a target machine that is separate and distinct from the workstation development machine. The principal target environment we'll be working within this book is an ARM-based single board computer introduced in Chapter 5. This is a widely available, relatively inexpensive device with capabilities typical of a real-world embedded product. But even if you choose not to purchase the board, we'll introduce an approach to simulation that will allow you to play around with many of the basic concepts on your workstation.

Personal Biases

Like most computer users, for better or worse, I've spent years in front of a Windows screen and, truth be told, I still use Windows for my day-to-day computing. But before that

I was perfectly at home with DOS and even before that I hacked away at RT-11, RSX-11, and VMS. So it's not like I don't understand command line programming. In fact, back in the pre-Windows 95 era, it was probably a couple of years before I finally added WIN to my AUTOEXEC.BAT file.

Hardcore Unix programmers, however, think GUIs are for wimps. They proudly do everything from the command line. Say what you will, but I like GUIs. Yes, the command line still has its place, particularly for shell scripts and makefiles, but for moving around the file hierarchy and doing simple file operations like move, copy, delete, rename, etc., drag-and-drop beats the miserably obscure Unix shell commands hands down. I also refuse to touch text-based editors like vi and emacs, although recently I've come to tolerate vim. Sure they're powerful if you can remember all those obscure commands. Give me a WYSIWYG editor any day.

My favorite GUI is the KDE desktop environment. It has all the necessary bells and whistles including a very nice syntax coloring editor, not to mention a complete suite of office and personal productivity tools. KDE, an acronym for "K Desktop Environment" by the way, is included in most commercial Linux distributions. Clearly, you're free to use whatever environment you're most comfortable with to work the book's examples. But if you're new to Linux, I would recommend KDE and that's the environment I'll be demonstrating as we proceed through the book.

Organization

The book is organized around three major sections.

Part 1—Chapters 1 through 6. It's an introduction and getting started guide. You'll install Linux if you don't already have it. There's an introduction to Linux itself for those not familiar with Unix/Linux systems. You'll set up and connect the target board that will be used in Part 2 to investigate cross-platform application development. Finally, you'll install and get familiar with various cross-development tools including Eclipse, the Open Source integrated development environment.

Part 2—Chapters 7 through 12—explores application programming in a cross-platform environment. We'll look at accessing hardware from user space, debugging techniques, and high-level simulation for preliminary debugging. We'll introduce Posix threads, network programming, and Android, and we'll briefly look at kernel-level programming and device drivers.

Part 3 goes on to look at various components and tools in the embedded Linux programmer's toolkit. We'll look at the U-boot boot loader, BusyBox, Buildroot, and

OpenEmbedded among others. Finally, we'll wrap up with an exploration of real-time enhancements to Linux.

OK, let's get on with it. Join me for a thrill-packed, sometimes bumpy, but ultimately fun and rewarding, roller-coaster ride through the exciting world of embedded Linux.

## PART 1

Introduction and Getting Started

## CHAPTER 1

The Embedded and Real-Time Space

*If you want to travel around the world and be invited to speak
at a lot of different places, just write a Unix operating system.*

Linus Torvalds

What Is Embedded?

You're at a party when an attractive member of the opposite sex approaches and asks you what you do. You could be flip and say something like "as little as possible," but eventually the conversation will get around to the fact that you write software for embedded systems. Before your new acquaintance starts scanning around the room for a lawyer or doctor to talk to, you'd better come up with a captivating explanation of just what the heck embedded systems are.

I usually start by saying that an embedded system is a device that has a computer inside it, but the user of the device doesn't necessarily know, or care, that the computer is there. It's hidden. The example I usually give is the engine control computer in your car. You don't drive the car any differently because the engine happens to be controlled by a computer. Oh, and there's a computer that controls the antilock brakes, another to decide when to deploy the airbags, and any number of additional computers that keep you entertained and informed as you sit in the morning's bumper-to-bumper traffic. In fact, the typical car today has more raw computing power than the Lunar Lander of the 1970s. Heck, your cell phone probably has more computing power than the Lunar Lander.

You can then go on to point out that there are a lot more embedded computers out in the world than there are personal computers (PCs).[1] In fact, recent market data shows that PCs account for only about 2% of the microprocessor chips sold every year. The average house contains at least a couple dozen computers even if it doesn't have a PC.

From the viewpoint of programming, embedded systems show a number of significant differences from conventional "desktop" applications. For example, most desktop applications deal with a fairly predictable set of input/output (I/O) devices—a disk, graphic

---

[1] In today's world though, most people are at least vaguely aware of embedded computing devices because they carry at least one, their cell phone.

display, a keyboard, mouse, sound card, and network interface. And these devices are generally well supported by the operating system (OS). The application programmer doesn't need to pay much attention to them.

Embedded systems on the other hand often incorporate a much wider variety of I/O devices than typical desktop computers. A typical system may include user I/O in the form of switches, pushbuttons, and various types of displays often augmented with touch screens. It may have one or more communication channels, either asynchronous serial, Universal Serial Bus (USB), and/or network ports. It may implement data acquisition and control in the form of analog-to-digital (A/D) and digital-to-analog (D/A) converters. These devices seldom have the kind of OS support that application programmers are accustomed to. Therefore, the embedded systems programmer often has to deal directly with the hardware.

Embedded devices are often severely resource-constrained. Although a typical PC now has 4 GB of RAM and several hundred gigabytes of disk, embedded devices often get by with a few megabytes of RAM and non-volatile storage. This too requires creativity on the part of the programmer.

What Is Real-Time?

Real-time is even harder to explain. The basic idea behind real-time is that we expect the computer to respond to its environment *in time*. But what does "in time" mean? Many people assume that real-time means real fast. Not true. Real-time simply means *fast enough* in the context in which the system is operating. If we're talking about the computer that runs your car's engine, that's fast! That guy has to make decisions—about fuel flow, spark timing—every time the engine makes a revolution.

On the other hand, consider a chemical refinery controlled by one or more computers. The computer system is responsible for controlling the process and detecting potentially destructive malfunctions. But chemical processes have a time constant in the range of seconds to minutes at the very least. So we would assume that the computer system should be able to respond to any malfunction in sufficient time to avoid a catastrophe.

But suppose the computer were in the midst of printing an extensive report about last week's production or running payroll when the malfunction occurred. How soon would it be able to respond to the potential emergency?

The essence of real-time computing is not only that the computer responds to its environment fast enough but that it also responds *reliably* fast enough. The engine control computer must be able to adjust fuel flow and spark timing every time the engine turns over. If it's late, the engine doesn't perform right. The controller of a chemical plant must be able to detect and respond to abnormal conditions in sufficient time to avoid a catastrophe. If it doesn't, it has failed.

I think this quote says it best:

> *A real-time system is one in which the correctness of the computations not only depends upon the logical correctness of the computation but also upon the time at which the result is produced. If the timing constraints of the system are not met, system failure is said to have occurred.*
>
> **Donald Gillies in the Real-Time Computing FAQ**

So the art of real-time programming is designing systems that reliably meet timing constraints in the midst of random asynchronous events. Not surprisingly this is easier said than done, and there is an extensive body of literature and development work devoted to the theory of real-time systems.

How and Why Does Linux Fit In?

Linux developed as a general-purpose OS in the model of Unix whose basic architecture it emulates. No one would suggest that Unix is suitable as an embedded or real-time OS (RTOS). It's big, it's a resource hog and its scheduler is based on "fairness" rather than priority. In short, it's the exact antithesis of an embedded OS.

But Linux has several things going for it that earlier versions of Unix lack. It's "free" and you get the source code. There is a large and enthusiastic community of Linux developers and users. There's a good chance that someone else either is working or has worked on the same problem you're facing. It's all out there on the web. The trick is finding it.

Open Source

Linux has been developed under the philosophy of Open Source software pioneered by the Free Software Foundation (FSF). Quite simply, Open Source is based on the notion that software should be freely available: to use, to modify, and to copy. The idea has been around for some 20 years in the technical culture that built the Internet and the World Wide Web and in recent years has spread to the commercial world.

There are a number of misconceptions about the nature of Open Source software. Perhaps the best way to explain what it is, is to start by talking about what it isn't.

- Open Source is not shareware. A precondition for the use of shareware is that you pay the copyright holder a fee. Shareware is often distributed in a free form that is either time- or feature-limited. To get the full package, you have to pay. By contrast, Open Source code is freely available and there is no obligation to pay for it.

- Open Source is not public domain. Public domain code, by definition, is not copyrighted. Open Source code is copyrighted by its author who has released it under the terms of an Open Source software license. The copyright owner thus gives you the right to use the code provided you adhere to the terms of the license. But if you don't comply with the terms of the license, the copyright owner can demand that you stop using the code.
- Open Source is not necessarily free of charge. Having said that there's no obligation to pay for Open Source software doesn't preclude you from charging a fee to package and distribute it. A number of companies are in the specific business of selling packaged "distributions" of Linux.

Why would you pay someone for something you can get for free? Presumably, because everything is in one place and you can get some support from the vendor. Of course, the quality of support greatly depends on the vendor.

So "free" refers to freedom to use the code and not necessarily zero cost. Think "free speech," not "free beer."

Open Source code is:

- Subject to the terms of an Open Source license, in many cases the GNU General Public License (GPL) (see below).[2]
- *Subject to critical peer review.* As an Open Source programmer, your code is out there for everyone to see and the Open Source community tends to be a very critical group. Open Source code is subject to extensive testing and peer review. It's a Darwinian process in which only the best code survives. "Best" of course is a subjective term. It may be the best *technical* solution but it may also be completely unreadable.
- *Highly subversive.* The Open Source movement subverts the dominant paradigm, which says that intellectual property such as software must be jealously guarded so you can make a lot of money off of it. In contrast, the Open Source philosophy is that software should be freely available to everyone for the maximum benefit of society. And if you can make some money off of it, that's great, but it's not the primary motivation. Richard Stallman, founder of the FSF, is particularly vocal in advocating that software should not have owners (see Appendix C).

Not surprisingly, Microsoft sees Open Source as a serious threat to its business model. Microsoft representatives have gone so far as to characterize Open Source as "un-American." On the other hand, many leading vendors of Open Source software give their programmers and engineers company time to contribute to the Open Source community. And it's not just charity, it's good business!

---

[2] GNU is a 'self-referential acronym' that means 'GNU's not Unix'. The GNU Project, initiated by Richard Stallman in 1983, aims for a 'complete Unix-compatible software system' composed entirely of free software.

Portable and Scalable

Linux was originally developed for the Intel x86 family of processors and most of the ongoing kernel development work continues to be on x86s. Nevertheless, the design of the Linux kernel makes a clear distinction between processor-dependent code, which must be modified for each different architecture, and code that can be ported to a new processor simply by recompiling it. Consequently, Linux has been ported to a wide range of 32-bit, and more recently 64-bit, processor architectures including:

- Motorola 68k and its many variants
- Alpha
- Power PC
- Advanced RISC Machines (ARM)
- Sparc
- MIPS

to name a few of the more popular. So whatever 32-bit architecture you're considering for your embedded project, chances are there's a Linux port available for it and a community of developers supporting it.

A typical desktop Linux installation runs into 10−20 GB of disk space and requires a gigabyte of RAM to execute decently. By contrast, embedded targets are often limited to 64 MB or less of RAM and perhaps 128 MB of flash ROM for storage. Fortunately, Linux is highly modular. Much of that 10 GB represents documentation, desktop utilities, and options like games that simply aren't necessary in an embedded target. It is not difficult to produce a fully functional, if limited, Linux system occupying no more than 2 MB of flash memory.

The kernel itself is highly configurable and includes reasonably user-friendly tools that allow you to remove kernel functionality not required in your application.

Where Is Linux Embedded?

Just about everywhere. As of July 2005, the web site *LinuxDevices.com* listed a little over 300 commercially available products running Linux. They range from cell phones, Personal Digital Assistant (PDAs), and other handheld devices through routers and gateways, thin clients, multimedia devices, and TV set-top boxes to robots and even ruggedized VMEbus,[3] chassis suitable for military command and control applications. And these are just the products the *LinuxDevices* editors happen to know about.

One of the first and perhaps the best-known home entertainment devices to embed Linux is the TiVo Personal Video Recorder that created a revolution in television viewing when it

---

[3] VERSA Module Eurocard bus, an industrial computer packaging and interconnection standard.

was first introduced in 2000. The TiVo is based on a power PC processor and runs a "home grown" embedded Linux port that uses a graphics rendering chip for generating video.

Half the fun of having a device that runs Linux is making it do something more, or different, than what the original manufacturer intended. There are a number of web sites and books devoted to hacking the TiVo. Increasing the storage capacity is perhaps the most obvious hack. Other popular hacks include displaying weather, sports scores, or stock quotes, and setting up a web server.

Applications for embedded Linux aren't limited to consumer products. It's found in point of sale terminals, video surveillance systems, robots, even in outer space. NASA's Goddard Space Flight Center developed a version of Linux called FlightLinux to address the unique problems of spacecraft's onboard computers. On the International Space Station, Linux-based devices control the rendezvous and docking operations for unmanned servicing spacecraft called Automatic Transfer Vehicles.

Historically, telecommunications carriers and service providers have relied on specialized, proprietary platforms to meet the availability, reliability, performance, and service response time requirements of telecommunication networks. Today, carriers and service providers are embracing "open architecture" and moving to commercial off-the-shelf hardware and software in an effort to drive down costs while still maintaining carrier-class performance.

Linux plays a major role in the move to open, standards-based network infrastructure. In 2002, the Open Source Development Lab set up a working group to define "Carrier Grade Linux" (CGL) in an effort to meet the higher availability, serviceability, and scalability requirements of the telecom industry. The objective of CGL is to achieve a level of reliability known as "five nines," meaning the system is operational 99.999% of the time. That translates into no more than about 5 min of downtime in a year.

Open Source Licensing

Most End-User License Agreements for software are specifically designed to restrict what you are allowed to do with the software covered by the license. Typical restrictions prevent you from making copies or otherwise redistributing it. You are often admonished not to attempt to "reverse-engineer" the software.

By contrast, an Open Source license is intended to guarantee your rights to use, modify, and copy the subject software as much as you'd like. Along with the rights comes an obligation. If you modify and subsequently distribute software covered by an Open Source license, you are obligated to make available the modified source code under the same terms. The changes become a "derivative work" which is also subject to the terms of the license.

This allows other users to understand the software better and to make further changes if they wish.

Open Source licenses are called "copyleft" licenses, a play on the word copyright intended to convey the idea of using copyright law as a way of enhancing access to intellectual property like software rather than restricting it. Whereas copyright is normally used by an author to prevent others from reproducing, adapting, or distributing a work, copyleft explicitly allows such adaption and redistribution provided the resulting work is released under the same license terms. Copyleft thus allows you to benefit from the work of others, but any modifications you make must be released under similar terms.

Arguably, the best-known, and most widely used, Open Source license is the GNU General Public License (GPL) first released by the FSF in 1989. The Linux kernel is licensed under the GPL. But the GPL has a problem that makes it unworkable in many commercial situations. Software that makes use of, or relies upon, other software released under the GPL, even just *linking* to a library, is considered a derivative work and is therefore subject to the terms of the GPL and must be made available in source code form.

To get around this, and thus promote the development of Open Source libraries, the FSF came up with the "Library GPL." The distinction is that a program linked to a library covered by the LGPL is not considered a derivative work and so there's no requirement to distribute the source, although you must still make available the source to the library itself.

Subsequently, the LGPL became known as the "Lesser GPL" because it offers less freedom to the user. So while the LGPL makes it possible to develop proprietary products using Open Source software, the FSF encourages developers to place their libraries under the GPL in the interest of maximizing openness.

At the other end of the scale is the Berkeley Software Distribution (BSD) license, which predates the GPL by some 12 years. It "suggests," but does not require, that source code modifications be returned to the developer community and it specifically allows derived products to use other licenses, including proprietary ones.

Other licenses—and there are quite a few—fall somewhere between these two poles. The Mozilla Public License (MPL), for example, developed in 1998 when Netscape made its browser open source, contains more requirements for derivative works than the BSD license, but fewer than the GPL or LGPL. The Open Source Initiative, a non-profit group that certifies licenses meeting its definition of Open Source, lists 78 certified licenses on its web site as of December 2011.

Most software released under the GPL, including the Linux kernel, is covered by version 2 of the license, which was released in 1991, coincidentally the same year Linux was born. FSF released version 3 of the GPL in June of 2007. One of the motivations for version 3 was to address the problem of "tivoization," a term coined by Richard Stallman, founder of the FSF. It turns out that TiVo will only run code with an authorized digital signature. So even though TiVo makes the source code available in compliance with the GPL, modifications to that code won't run.

Stallman considers this circumventing the spirit of the GPL. Other developers, including Linus Torvalds, see digital signatures as a useful security tool and wouldn't want to ban them outright. The debate continues. In any case, the kernel itself is unlikely to move up to version 3 anytime soon.

Legal Issues

Considerable FUD[4] has been generated about the legal implications of Open Source, particularly in light of SCO's claims that the Linux kernel is "contaminated" with its proprietary Unix code. The SCO Group, formerly known as Santa Cruz Operations, acquired the rights to the Unix System V source code from Novell in 1996, although there is some dispute as to exactly what SCO bought from Novell. In any case, SCO asserted that IBM introduced pieces of SCO's copyrighted, proprietary Unix code into the Linux kernel and is demanding license fees from Linux users as a consequence.

Ultimately, SCO's case collapsed and the company filed for Chapter 11 bankruptcy in 2007. But all of a sudden there is serious money to be made by fighting over Open Source licensing issues. The upshot is that embedded developers need to be aware of license issues surrounding both Open Source and proprietary software. Of necessity, embedded software is often intimately tied to the OS and includes elements derived or acquired from other sources. While no one expects embedded engineers to be intellectual property attorneys, it is nevertheless essential to understand the license terms of the software you use and create to be sure that all the elements "play nicely" together.

And the issue cuts both ways. There are also efforts to identify violations of the GPL. The intent here is not to make money but to defend the integrity of the GPL by putting pressure on violators to clean up their act. In particular, the GPL Violations Project has "outed" a dozen or so embedded Linux vendors who appear to have played fast and loose with the GPL terms. According to Harald Welte, founder of the GPL Violations Project, the most common offenders are networking devices such as routers, followed by set-top boxes, and vehicle navigation systems.

Open Source licensing expert Bruce Perens has observed that embedded developers seem to have a mindset that "this is embedded, no one can change the source—so the GPL must not really apply to us." It does.

---

[4] Fear, Uncertainty, and Doubt.

Now that we have some idea of the embedded and real-time space and how Linux fits into it, Chapter 2 describes the process of installing Linux on a workstation.

Resources

embedded.com—The web site for *Embedded Systems Design* magazine. This site is not specifically oriented to Linux but is quite useful as a more general information tool for embedded system issues.

fsf.org—The Free Software Foundation.

gpl-violations.org—the General Public License Violations Project was started to "raise the awareness about past and present violations" of the General Public License. According to the web site, it is "still almost a one-man effort."

keegan.org/jeff/tivo/hackingtivo.html—One of the many sites devoted to TiVo hacks.

kernel.org—The Linux kernel archive. This is where you can download the latest kernel versions as well as virtually any previous version.

Linux resources on the Web are extensive. This is a list of some sites that are of particular interest to embedded developers.

linuxdevices.com—A news and portal site devoted to the entire range of issues surrounding embedded Linux.

opensource.org—The Open Source Initiative (OSI), a non-profit corporation "dedicated to managing and promoting the Open Source Definition for the good of the community." OSI certifies software licenses that meet its definition of Open Source.

osdl.org—Open Source Development Lab (OSDL), a non-profit consortium focused on accelerating the growth and adoption of Linux in both the enterprise and, more recently, embedded spaces. In September of 2005, OSDL took over the work of the Embedded Linux Consortium, which had developed a "platform specification" for embedded Linux.

sourceforge.net—"World's largest Open Source development web site" provides free services to Open Source developers including project hosting and management, version control, bug and issue tracking, backups and archives, and communication and collaboration resources.

slashdot.org—"News for nerds, stuff that matters." A very popular news and forum site focused on Open Source software in general and Linux in particular.

uclinux.org—The Linux/Microcontroller project is a port of Linux to systems without a Memory Management Unit.

## CHAPTER 2

Installing Linux

If Bill Gates had a nickel for every time Windows crashed... Oh wait, he does.
Spotted on Slashdot.org

While it is possible to do embedded Linux development under Windows, it's not easy. Basic Windows lacks many of the tools and features that facilitate embedded development. And why would you want to do that anyway? So we'll keep our focus firmly on Linux and not delve into the process of setting up a Windows development system.

Even though this book is not an introduction to Linux, it's worth taking a little time to review the installation process and alternative configurations. If you already have a Linux installation that you're happy with, you can probably skip this chapter unless you want to learn about virtualization or dual-booting.

The instructions and steps in this chapter are primarily oriented toward Fedora, but the general concepts apply pretty much to any Linux installation.

Distributions

Linux installation has improved substantially over the years to the point that it's a reasonably straightforward, just about painless, process. The easiest, and indeed the only sensible, way to install Linux is from a *distribution* or "distro" for short. There are probably several hundred distributions floating around on the Internet, many of which address specific niches such as embedded.

A distribution contains just about everything you would need in your Linux installation including the kernel, utility programs, graphical desktop environments, development tools, games, media players, and on and on. Most, but not all, distributions make use of a *package manager*, a software that combines the individual components into easily managed packages. Fedora and Red Hat distributions use RPM, the Red Hat Package Manager. Debian and its derivatives such as Ubuntu use dpkg. A package contains the software for the component, i.e., the executables, libraries, and so on, plus scripts for installing and uninstalling the component and dependency information showing what other packages this one requires.

A typical distro may incorporate several thousand packages. In fact, this becomes a marketing game among the various projects and distributors of Linux. "My distro has more packages than your distro." Here are some of the more popular and user-friendly

Linux for Embedded and Real-time Applications.
© 2013 Elsevier Inc. All rights reserved.

13

DOI: http://dx.doi.org/10.1016/B978-0-12-415996-9.00002-2

distributions, any of which are suitable for embedded development. Web locations are given in the Resources section at the end of this chapter.

Debian GNU/Linux

The Debian Project is an all-volunteer effort comprising some 3,000 individuals around the world supported entirely by donations. A 2007 survey by SurveyMonkey.com showed Debian to be second only to Ubuntu, itself a Debian derivative, as "most used Linux distribution for both personal and organizational use."

Debian is known for its "abundance of options." The current stable release, 6.0, has over 29,000 packages supporting nine computer architectures including the usual Intel/AMD 32- and 64-bit processors as well as ARM and IBM eServer zSeries mainframes. Debian claims to be one of the first distributions to use a robust package management system, which is different from RPM. Debian packages get the file extension.deb.

Debian tends to have a longer and perhaps less-consistent release cycle than many of the other popular distributions typically taking 2 years between major releases.

Fedora

The Fedora Project was started in late 2003 when Red Hat discontinued its retail line of Linux distributions to focus on enterprise software. The original Fedora release was based on Red Hat Linux 9. Red Hat continues to sponsor the Fedora Project, which serves as a launching point for releases of Red Hat Enterprise Linux (RHEL).

Fedora prides itself on being on the leading edge of open source software development and, as a result, it has a relatively aggressive release schedule, about every 6 months. The current release is 16. Having started with Red Hat, Fedora continues to be my favorite Linux distribution and I'm currently running 14. I tend to stay one or two major releases behind the leading edge and only update about every two or three releases. I don't have to have the absolute latest and greatest.[1] Hey, if it works, stick with it.

I have two basic complaints about new releases. This may also apply to distros other than Fedora:

1. Features tend to "move around." I'm perhaps sensitive to this because I'm inclined to use the graphical dialogs for configuration and setup. It seems like with each new release, the path to these dialogs changes.

---

[1] And neither do you, by the way. The basic functionality doesn't change that much from one release to the next. Much of it is cosmetic. If you're just starting out, pick a fairly recent release and stick with it for a while.

2. Each successive release seems to get more "Windows-like." That is, more menus turn into cutsie icons rather than simple, straightforward text selections. This no doubt reflects a desire to appeal to a more "consumer" audience, but as a computer professional, I find it off-putting. Fortunately, at least for now, there are ways to revert back to the classical menu styles.

While just about everything in this book is "distribution-agnostic," one area where distros tend to differ is in setup and configuration. We'll confront that issue in Chapter 4 when we need to make some changes to networking. The descriptions there will emphasize Fedora. With that in mind, if you haven't yet settled on a favorite Linux distribution, I would recommend Fedora, at least while you're working through the book.

Fedora only runs on Intel/AMD platforms.

Red Hat Enterprise Linux

RHEL is specifically targeted at the commercial market, including mainframes. There are server versions for x86, x86-64, Itanium, PowerPC, and IBM System z, and desktop versions for x86 and x86-64. While RHEL derives many of its features from Fedora, it has a more conservative release schedule with major releases appearing about every 2 years. As of late 2011, the latest stable release is 6.1.

Even though RHEL is a commercial product, it is based entirely on Open Source code. Consequently, Red Hat makes the entire source code base available for download. Several groups have taken advantage of this to rebuild their own versions of RHEL. One of the best known of these is CentOS, said to be the eighth most popular distribution as of late 2011. These rebuilds remove any reference to Red Hat's trademarks and point the update systems at non-Red Hat servers. Otherwise, they are functionally identical.

While the rebuilds are free, they are of course not eligible for any kind of Red Hat support.

SUSE

SuSE was originally developed in Germany with the initial release in 1994 making it the oldest existing commercial distribution. The name is a German acronym for Software und System Entwicklung (Software and Systems Development). The name was subsequently changed to SUSE and is no longer considered an acronym.

Novell acquired SUSE Linux AG in 2003 and in 2005 announced the openSUSE project to allow outside developers to participate. SUSE Linux is available in two forms: openSUSE, driven by the openSUSE project, and SUSE Linux Enterprise, a commercial version. Much like Fedora and Red Hat, openSUSE is on the "bleeding edge" of Linux development with an aggressive release schedule while SUSE Linux Enterprise sticks to a more conservative schedule.

Ubuntu

According to its web site, Ubuntu is an ancient African word meaning "humanity to others." It also means "I am what I am because of who we all are." It is a fork of the Debian code base first released in 2004 with the aim of creating an easy-to-use version of Linux. It's available in two versions: desktop and server.

Releases are made on a predictable 6 month schedule with each fourth release getting long-term support (LTS). LTS releases have been supported for 3 years for the desktop version and 5 years for the server version. With the pending release of version 12.04, desktop support will be extended to 5 years.

Historically, Ubuntu, like most other Linux distros, supported the graphical desktop environments, both GNU Object Model Environment (GNOME) and K Desktop Environment (KDE). Release 11.04 in spring 2011 introduced a new desktop environment called Unity that is a "shell interface" for GNOME. The intention is to make more efficient use of space on the limited size screens of notebooks and tablets. Some users have criticized the new interface as "too different from and less capable than GNOME" while others find the minimalist approach more appealing than the older paradigm.

While Ubuntu emphasizes ease of use, the developers have made basic system configuration more difficult by eliminating many of the graphical dialogs that support setup and configuration in, for example, Fedora. Furthermore, the root user account is "locked" so that it's not possible to directly log in as root. You must use the sudo command. This apparently is intentional, in the tradition of Windows, as a way of discouraging "average users" from messing around with their systems.

Difficulty in configuration will become an issue in Chapter 4 when we have to change some network parameters. I don't recommend Ubuntu unless it's your favorite distribution and you're very comfortable with it.

Notwithstanding, Ubuntu may well be the most popular Linux distribution. In June 2009, ZDNet reported "Worldwide, there are 13 million active Ubuntu users with use growing faster than any other distribution."

Hardware Requirements

Having selected a distribution, you need something to run it on. Any modern PC will work just fine as a development host. Minimum requirements are: a Pentium class processor,

1 GB of RAM for graphical operation, and at least 20 GB of disk for a "workstation" class Linux installation. Of course, more RAM and disk are always better.

I do most of my Linux work on a "virtual" machine (more about that later) running under Windows 7 on a Dell laptop. The virtual machine gets 1 GB of RAM and has a 20 GB disk. I currently run Fedora 14.

You will need at least one asynchronous serial port. A USB to serial converter should work fine. You will also need a network interface. We'll use a combination of serial and network ports to communicate with the target as well as debug target code.

Installation Scenarios

The next decision then is *how* you want to install Linux. There are basically three installation scenarios.

Stand-Alone

This is the obvious choice if you can dedicate a machine to Linux. You will let the installation process format the entire disk.

Dual-Booting

In many cases though, you'll probably want to install Linux on a machine that already runs some variant of Windows. There are a couple of approaches to that. This section describes dual-booting, and the next one describes virtualization.

In the dual-boot scenario, you select at boot time which OS to boot. That OS takes full control of the machine. The Linux installation will replace the standard Windows boot loader with GRUB, the GRand Unified Boot loader. GRUB then offers the option of selecting the OS to boot as shown in Figure 2.1.

While historically, dual-booting has been the most popular installation scenario for average users, it's also the most complicated because it requires reconfiguring your hard disk. The most common case is you already have a version of Windows installed on the machine and you want to add Linux.

At this point, Windows probably occupies the entire disk and so you have to make space available for Linux. Fundamentally, this requires reducing the size of the disk partition holding Windows so you can create unallocated space for Linux. Remember, that you'll need on the order of 20 GB of disk space for Linux, so your disk must have at least that much contiguous free space. You can use the Windows defragmentation tool to put all free space at the "end" of the disk.

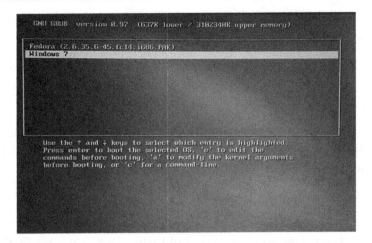

Figure 2.1
GRUB Boot Menu.

Windows 7 has a slick disk-management tool that handles partitioning. From the Start menu, select Control Panel > Administrative Tools > Computer Management. In the Computer Management dialog, select Storage > Disk Management. That displays the menu shown in Figure 2.2. This shows five partitions, the largest of which is identified as drive C:.

A little background on disk partitioning is in order here. In the DOS/Windows/PC world, a disk can have up to four *primary* partitions. Any of these primary partitions can be designated an *extended* partition, which in turn can hold several *logical* partitions. There's no fixed upper limit on the number of logical partitions an extended partition can hold, but owing to the way in which Linux accesses partitions, 12 is the practical upper limit on a single disk drive.

Take a look at the partition list in Figure 2.2 and compare it with the list in Listing 2.1 derived from the Linux fdisk command on the same disk. This particular machine is already configured for dual-booting. Note that fdisk reports six partitions whereas the Windows Disk Manager only shows five. /dev/sda4 is identified as an extended partition. sda5/ and sda6/ are logical partitions within sda4/. The Disk Manager doesn't show the extended partition.

Oddly, the Disk Manager identifies all five partitions as primary, but fdisk shows us that the two on the right of the graphic representation are in fact logical partitions within an extended partition.

With that background, it's time to create some empty space for Linux. But first, before you make *any* changes to a disk drive, **Always back up your data**. Even though these tools are

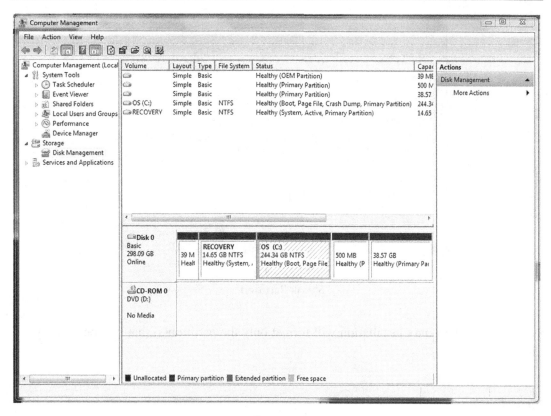

Figure 2.2
Windows 7 Disk Manager.

```
    Device Boot      Start          End      Blocks   Id   System
   /dev/sda1               63        80324        40131   de   Dell Utility
   /dev/sda2      *      81920     30801919     15360000    7   HPFS/NTFS
   /dev/sda3          30801920    543220399    256209240    7   HPFS/NTFS
   /dev/sda4         543221760    625141759     40960000    5   Extended
   /dev/sda5         543223808    544247807       512000   83   Linux
   /dev/sda6         544249856    625141759     40445952   8e   Linux LVM
```

Listing 2.1
fdisk Output.

supposed to work, stuff happens. So be prepared. Right-click on the OS (C:) partition and select Shrink volume…. Note, incidentally that Windows tends to call partitions volumes. The Disk Manager queries the partition to see how small it can be made. This can take a while on a large disk. Finally, the dialog of Figure 2.3 appears. While the labels are a little confusing, this is telling us that 113 MB can be removed from the C: partition. Clearly, you want to leave some space free on the C: drive, so you wouldn't want to take all 113 MB.

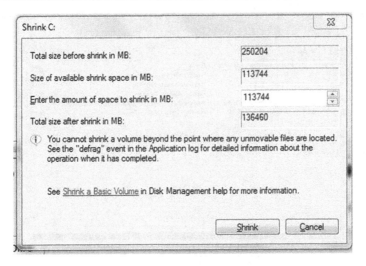

Figure 2.3
Shrink Partition.

Subsequently, the Linux installation will format only the partition(s) that will be used by Linux.

Virtualization

But the big buzzword in computing these days is virtualization, the process of running one OS on top of another. The base, or native, OS is called the *host*. It runs a *virtual machine manager*, VMM, that in turn runs one or more virtual machines called *guests*. Whereas in the dual-boot scenario one or the other OS is running exclusively, with virtualization the host and the guests are running simultaneously. You can even move files and data between them seamlessly.

There are two popular VMMs—VMware and VirtualBox. VMware is a commercial product from a company of the same name. They offer a free version called VMware Player that runs on both Windows and Linux hosts.

VirtualBox is an open source package sponsored by Oracle (formerly Sun Microsystems). It too is available for both Windows and Linux hosts, but also supports Mac OS X and Solaris.

The two packages are fairly similar in their installation and operation. After installing the software (see Resources section for download pages), you create one or more guest machines allocating resources to them such as disk and RAM. A "wizard" steps you through the process of creating a new machine. A disk in the guest machine is represented by a very large file in the host. Then you install an OS on the guest in much the same way you would install it natively.

I happen to use VirtualBox.

There's a third VMM called Parallels from a Swiss company of the same name. It is strictly a commercial product oriented towards graphics-intensive applications. You can get a free limited-time evaluation license. Parallels makes use of virtualization technology built into high-end Intel processors.

DVD or Live CD?

Most Linux distributions are available in at least three forms: a DVD, a collection of CDs (up to about six these days), or a *Live CD*. A Live CD is a minimal bootable Linux system that gives you an opportunity to play around with a new distribution without actually installing it. Of course, you can't save any data or modify the configuration. Every time it boots it reverts to the default configuration.

One of the options the Live CD offers is to install the distribution. When you select this option, it runs through essentially the same installation process as you would with a DVD or complete set of CDs. The difference is that packages to install are pulled from the network rather than from physical media.

Installation Process

Regardless of which scenario or installation medium you choose, the installation process is pretty much the same. This section describes Fedora installation, but other distributions have a very similar process:

- Download the distribution medium (see Resources section). This will be a .iso file, an exact image of either a DVD or a Live CD.
- Burn the .iso to the corresponding physical medium. You'll need a disk-burning program to transfer the .iso file(s) to physical media. In a virtual machine environment, you can mount the .iso file directly as if it were a CD or a DVD.
- Boot the computer from the installation medium or click the install icon from a Live CD. You may have to configure your PC's BIOS to boot from optical media.
- Follow the instructions.

The installation process itself is fairly straightforward. You are asked to select a number of common options such as language, keyboard, time zone, and so on. You'll select a root password. The root user is the system administrator, also known as the "superuser." More about that in the next chapter.

Disk Partitioning

Finally, you get to the dialog shown in Figure 2.4. This is where things get serious because you're about to make some irreversible decisions. If you're installing a Linux-only system or a virtual machine, the correct answer here is Use All Space. In the virtual case, "all space" really refers to the file that was created when you created the virtual machine. In the Linux-only case, it really is all the space on the disk. If you're installing a dual-boot configuration, the correct answer is Use Free Space, which is the space you freed up earlier in the Windows Disk Manager.

Unless you have good reason to do otherwise, accept the suggested partitioning. It generally works quite well. When you click Next you are prompted to confirm that you want to write changes to the disk. Up to this point the installer has made no permanent changes to your system. When you click Write changes to disk, you're committed.

Next, you'll be asked about the boot loader, GRUB. If you're installing a dual-boot, you have the option of selecting which OS is the default.

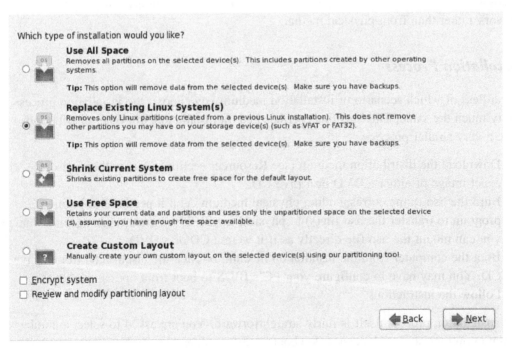

Figure 2.4
Disk Partitioning.

Package Selection

The next and final step before installation proceeds for real is package selection. In this menu, select Software Development. That preselects a number of packages that support software development including the GNU compiler collection (GCC). Select Customize now. When you click Next you'll see the dialog in Figure 2.5. This gives you the option of selecting additional features and packages beyond the Software Development group.

Under Desktop Environments, I recommend selecting KDE. This is very much a personal decision. There are two popular graphical desktop environments in the Linux world: GNOME and KDE. GNOME tends to be more popular. I happen to prefer KDE and, to the extent that the remainder of the book displays graphical desktop images, they will be KDE. I just happen to like the KDE presentation better. It's more "Windows-like." The truth is, I do most of my day-to-day computing work in Windows (hey, I'm writing this book in Word). If you've spent much of your career in front of a Windows screen, I think you'll find KDE more to your liking. But if you have a strong preference for GNOME, by all means, go ahead.

Under Development, select KDE Development. This installs the Qtopia (QT) graphic libraries that we'll need when we configure the Linux kernel in a few chapters.

Click Next and the installation begins. Not surprisingly, this will take a while, on the order of an hour, as something like a thousand packages will be installed on your computer.

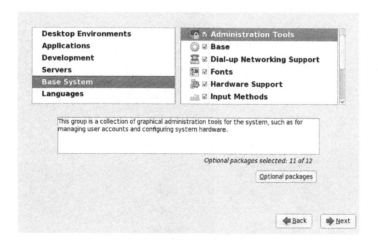

Figure 2.5
Package Selection.

When the installation is complete, you will be prompted to reboot. The system should reboot into what you just installed. There are a few post-installation issues to take care of. Among other things, you must accept the Fedora license terms and create a normal user.

Following that you'll have a fully functional Linux installation. Reboot it when you're ready for the next chapter where we'll review many of the useful features of Linux.

Resources

linuxiso.org—This site has downloadable ISO image files for virtually all of the popular Linux distributions on several architectures at various revision levels.

Specific Distribution Sites

debian.org
fedoraproject.org
opensuse.org
redhat.com
ubuntu.com

Other Resources

parallels.com—Site for the Parallels VMM.
virtualbox.org—Site for the VirtualBox VMM.
vmware.com—Information about the VMware virtual machine manager. Download VMware Player here.

## CHAPTER 3

Introducing Linux

There are two major products to come out of Berkeley:
LSD and Unix. We don't believe this to be a coincidence

Jeremy S. Anderson

For those who may be new to Unix-style OSs, this chapter provides an introduction to some of the salient features of Linux, especially of interest to embedded developers. This is by no means a thorough introduction, and there are many books available that delve into these topics in much greater detail.

Feel free to skim, or skip this chapter entirely, if you are already comfortable with Unix and Linux concepts.

Running Linux—KDE

Boot up your Linux machine and log in as your normal user. If you're running the KDE on Fedora 14, you'll see the screen shown in Figure 3.1. Unless you're running VirtualBox or VMware, you won't see the top two lines and the very bottom menu line. If you're running a different version of Linux, your screen will probably look different but should have most of the same features.

At the bottom left is a menu icon looking like a stylized "f" that serves the same purpose as the Start Menu in Windows. KDE calls this the "Application Launcher Menu." Initially, clicking this icon brings up a set of cutsie icon-based menus that I personally find difficult to use. Fortunately, you can change it to the older list-based style. Right-click the "f" icon and select `Switch to Classic Menu Style`. Recognize of course that this is purely a matter of personal taste.

File Manager

One of the first things I want to do with a new system is open a file manager so I can see what's on the system. Click the `Home` icon in the `Desktop Folder`. The file manager displays the contents of your *home directory*, the place where you store all of your own files. Note that by default KDE uses a single click to activate items. You can change that to double click by selecting `Settings>System Settings` from the Start Menu. Scroll down to `Hardware` and select `Input Devices>Mouse`.

Figure 3.1
Initial KDE Screen.

The default file manager in recent Fedora releases is called Dolphin and, again, is not my cup of tea. I find the older file manager, Konqueror, to be easier to use and provide a more useful presentation. You can change file managers from the `Default Applications` menu under System Settings.

Figure 3.2 shows Konqueror as I have it configured. Not surprisingly, that's not how it looks initially. To get the navigation panel on the left, click on the red "Root folder" icon on the far left. There are lots of options in Konqueror, so play around with it to get it exactly to your liking.

Shell Window

The other window that you'll be using alot is the command shell that we'll describe later in the chapter. From the Application Launcher Menu, select `System>Konsole (Terminal)`. Figure 3.3 shows how I have the shell configured. Again, there are numerous configuration options that are accessed by selecting `Configure Profiles...` from the `Settings` menu. I happen to like black text on a white background and I set the size to 80 × 25 because that matches the old serial CRT terminals that I used when I was starting out in this business.

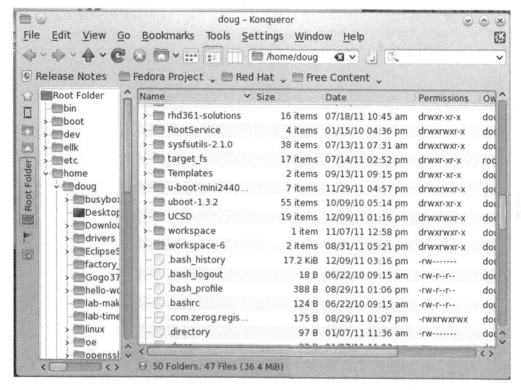

Figure 3.2
Konqueror File Manager.

The `File` menu offers options to open new shell windows and multiple tabs within a shell window. It is not unusual to have two or three shell windows open simultaneously and have two or three tabs in any of those windows.

Linux Features

Here are some of the important features of Linux and Unix-style OSs in general.

- *Multitasking.* The Linux scheduler implements true, preemptive multitasking in the sense that a higher priority process made ready by the occurrence of an asynchronous event will preempt the currently running process. But while it is preemptible,[1] there are relatively large latencies in the kernel that make it generally unsuitable for hard

---

[1] Is it "preemptible" or "preemptable"? Word 2007s spelling checker says they're both wrong. A debate on linuxdevices.com a while back seemed to come down on the side of "ible" but not conclusively. I think I'll stick with preemptible.

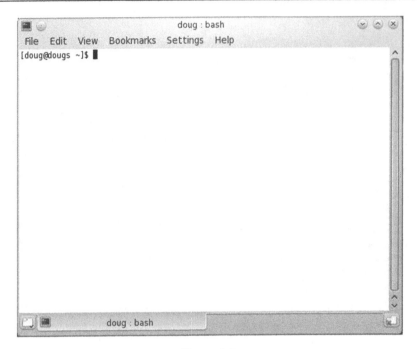

Figure 3.3
Command Shell Window.

real-time applications. Additionally, the default scheduler implements a "fairness" policy that gives all processes a chance to run.

- *Multiuser.* Unix evolved as a time-sharing system that allowed multiple users to share an expensive (at that time anyway) computer. Thus, there are a number of features that support privacy and data protection. Linux preserves this heritage and puts it to good use in server environments.[2]

- *Multiprocessing.* Linux offers extensive support for true symmetric multiprocessing where multiple processors are tightly coupled through a shared memory bus. This has become particularly significant in the era of multicore processors.

- *Protected Memory.* Each Linux process operates in its own private memory space and is not allowed to directly access the memory space of another process. This prevents a wild pointer in one process from damaging the memory space of another process. The errant access is trapped by the processor's memory protection hardware, and the process is terminated with appropriate notification.

---

[2] Although my experience in the embedded space is that the protection features, particularly file permissions, can be a downright nuisance. Some programs, the Firefox browser is an example, don't have the courtesy to tell you that you don't have permission to write the file, they just sit there and do nothing.

- *Hierarchical File System.* Yes, all modern OSs—even DOS—have hierarchical file systems. But the Linux/Unix model adds a couple of nice wrinkles on top of what we're used to with traditional PC OSs:
  - *Links.* A link is simply a file system entry that points to another file rather than being a file itself. Links can be a useful abstraction mechanism and a way to share files among multiple users. They also find extensive use in configuration scenarios for selecting one of several optional files.
  - *Device-Independent I/O.* Again, this is nothing new, but Linux takes the concept to its logical conclusion by treating every peripheral device as an entry in the file system. From an application's viewpoint, there is absolutely no difference between writing to a file and writing to, say, a printer.

Protected Mode Architecture

The implementation of protected mode memory in contemporary Intel processors first made its appearance in the 80386. It utilizes a full 32-bit address for an addressable range of 4 GB. Access is controlled such that a block of memory may be Executable, Read only, or Read/Write.

The processor can operate in one of four *privilege levels*. A program running at the highest privilege level, level 0, can do anything it wants—execute I/O instructions, enable and disable interrupts, and modify descriptor tables. Lower privilege levels prevent programs from performing operations that might be "dangerous." A word-processing application probably shouldn't be messing with interrupt flags, for example. That's the job of the OS.

So application code typically runs at the lowest level while the OS runs at the highest level. Device drivers and other services may run at the intermediate levels. In practice however, Linux and most other OSs for Intel processors only use levels 0 and 3. In Linux, level 0 is called "Kernel Space" while level 3 is called "User Space."

Real Mode

To begin our discussion of protected mode programming in the x86, it's useful to review how "real" address mode works.

Back in the late 1970s, when Intel was designing the 8086, the designers faced the dilemma of how to access a megabyte of address space with only 16 bits. At the time, a megabyte was considered an immense amount of memory. The solution they came up with, for better or worse, build a 20-bit (1 MB) address out of two 16-bit quantities called the Segment and Offset. Shifting the segment value four bits to the left and adding it to the offset creates the 20-bit linear address (Figure 3.4).

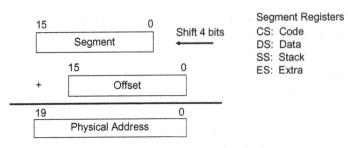

Figure 3.4
Real Mode Addressing.

The x86 processors have four segment registers in real mode. Every reference to memory derives its segment value from one of these registers. By default, instruction execution is relative to the Code Segment (CS), most data references (the MOV instruction for example) are relative to the Data Segment (DS), and instructions that reference the stack are relative to the Stack Segment (SS). The Extra Segment (ES) is used in string move instructions and can be used whenever an extra DS is needed. The default segment selection can be overridden with segment prefix instructions.

A segment can be up to 64 KB long and is aligned on 16-byte boundaries. Programs less than 64 KB are inherently position-independent and can be easily relocated anywhere in the 1 MB address space. Programs larger than 64 KB, either in code or data, require multiple segments and must explicitly manipulate the segment registers.

Protected Mode

Protected mode still makes use of the segment registers but instead of providing a piece of the address directly, the value in the segment register (now called the *selector*) becomes an index into a table of *Segment Descriptors*. The segment descriptor fully describes a block of memory including, among other things, its base and limit (Figure 3.5). The linear address in physical memory is computed by adding the offset in the logical address to the base contained in the descriptor. If the resulting address is greater than the limit specified in the descriptor, the processor signals a memory protection fault.

A descriptor is an 8-byte object that tells us everything we need to know about a block of memory.

Base Address [31:0]   Starting address for this block/segment.
Limit [19:0]   Length of this segment—This may be either the length in bytes (up to 1 MB) or the length in 4 KB pages. The interpretation is defined by the Granularity bit.
Type   A 4-bit field that defines the kind of memory that this segment describes.

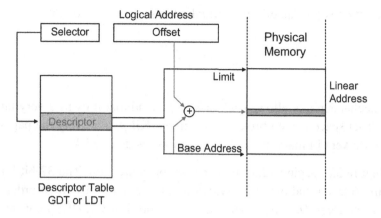

Figure 3.5
Protected Mode Addressing.

S   0 = This descriptor describes a "System" segment. 1 = This descriptor describes a CS or DS.

DPL   Descriptor Privilege Level—A 2-bit field that defines that minimum privilege level required to access this segment.

P   Present—1 = The block of memory represented by this descriptor is present in memory. Used in paging.

G   Granularity—0 = Interpret limit as bytes. 1 = Interpret Limit as 4 KB pages.

Note that with the Granularity bit set to 1, a single segment descriptor can represent the entire 4 GB address space.

Normal descriptors (S bit = 1) describe memory blocks representing data or code. The Type field is 4 bits where the most significant bit distinguishes between CS and DS. CSs are executable but DSs are not. A CS may or may not also be readable. A DS may be writable. Any attempted access that falls outside the scope of the Type field—attempting to execute a DS, for example—causes a memory protection fault.

"Flat" vs. Segmented Memory Models

Because a single descriptor can reference the full 4 GB address space, it is possible to build your system by reference to a single descriptor. This is known as "flat" model addressing and is, in effect, a 32-bit equivalent of the addressing model found in most 8-bit microcontrollers as well as the "tiny" memory model of DOS. All memory is equally accessible, and there is no protection.

Linux actually does something similar. It uses separate descriptors for the OS and each process so that protection is enforced, but it sets the base address of every descriptor to zero. Thus, the offset is the same as the virtual address. In effect, this does away with segmentation.

Paging

Paging is the mechanism that allows each task to pretend that it owns a very large flat address space. That space is then broken down into 4 KB *pages*. Only the pages currently being accessed are kept in main memory. The others reside on disk.

As shown in Figure 3.6, paging adds another level of indirection. The 32-bit linear address derived from the selector and offset is divided into three fields. The high-order 10 bits serve as an index into the *Page Directory*. The Page Directory Entry (PDE) points to a *Page Table*. The next 10 bits in the linear address provide an index into that table. The Page Table Entry (PTE) provides the base address of a 4 KB page in physical memory called a *Page Frame*. The low-order 12 bits of the original linear address supplies the offset into the page frame. Each task has its own Page Directory pointed to by processor control register CR3.

At either stage of this lookup process, it may turn out that either the Page Table or the Page Frame is not present in physical memory. This causes a *Page Fault*, which in turn causes the OS to find the corresponding page on disk and load it into an available page in memory. This in turn may require "swapping out" the page that currently occupies that memory.

A further advantage to paging is that it allows multiple tasks or processes to easily share code and data by simply mapping the appropriate sections of their individual address spaces into the same physical pages.

Paging is optional, you don't have to use it, although Linux does. Paging is controlled by a bit in processor register CR0.

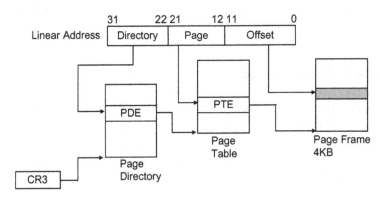

Figure 3.6
Paging.

Page Directory and PTEs are each 4 bytes long, so the Page Directory and Page Tables are a maximum of 4 KB, which also happens to be the Page Frame size. The high-order 20 bits point to the base of a Page Table or Page Frame. Bits 9–11 are available to the OS for its own use. Among other things, these could be used to indicate that a page is to be "locked" in memory, i.e., not swappable.

Of the remaining control bits the most interesting are as follows:

P   Present. 1 = This page is in memory. If this bit is 0, referencing this Page Directory or PTE causes a page fault. Note that when P = = 0 the remainder of the entry is not relevant.

A   Accessed. 1 = This page has been read or written. Set by the processor but cleared by the OS. By periodically clearing the Accessed bits, the OS can determine which pages haven't been referenced in a long time and are therefore subject to being swapped out.

D   Dirty. 1 = This page has been written. Set by the processor but cleared by the OS. If a page has not been written to, there is no need to write it back to disk when it has to be swapped out.

The Linux Process Model

The basic structural element in Linux is a *process* consisting of executable code and a collection of *resources* like data, file descriptors, and so on. These resources are fully protected such that one process can't directly access the resources of another. In order for two processes to communicate with each other, they must use the inter-process communication mechanisms defined by Linux such as shared memory regions or pipes.

This is all well and good as it establishes a high degree of protection in the system. An errant process will most likely be detected by the system and thrown out before it can do any damage to other processes (Figure 3.7). But there's a price to be paid in terms of excessive overhead in creating processes and using the inter-process communication mechanisms.

A *thread* however is code only. Threads only exist within the context of a process, and all threads in one process share its resources. Thus, all threads have equal access to data memory and file descriptors. This model is sometimes called *lightweight multitasking* to distinguish it from the Unix/Linux process model.

The advantage of lightweight tasking is that inter-thread communication is more efficient. The drawback of course is that any thread can clobber any other thread's data. Historically, most RTOSs have been structured around the lightweight model. In recent years of course, the cost of memory protection hardware has dropped dramatically. In response, many

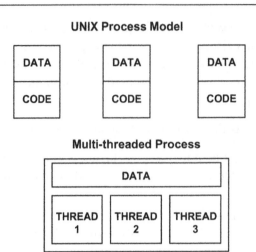

Figure 3.7
Processes vs. Threads.

RTOS vendors now offer protected mode versions of their systems that look like the Linux process model.

The fork() Function

Linux starts life with one process, the `init` process, created at boot time. Every other process in the system is created by invoking `fork()`. The process calling `fork()` is termed the *parent* and the newly created process is termed the *child*. So every process has ancestors and may have descendants depending on who created who.

If you've grown up with multitasking OSs where tasks are created from functions by calling a task creation service, the fork process can seem downright bizarre. `fork()` creates a *copy of the parent process*—code, data, file descriptors, and any other resources the parent may currently hold. This could add up to megabytes of memory space to be copied. To avoid copying a lot of stuff that may be overwritten anyway, Linux employs a *copy-on-write* strategy.

`fork()` begins by making a copy of the process data structure and giving it a new process identifier (PID) for the child process. Then, it makes a new copy of the Page Directory and Page Tables. Initially, the PTEs all point to the same physical pages as the parent process. All pages for both processes are set to read only. When one of the processes tries to write, that causes a page fault, which in turn causes Linux to allocate a new page for that process and copy over the contents of the existing page.

```
#include <unistd.h>
#include <>

pid_t pid;

void do_child_thing (void)
{
    printf ("I am the child. My PID is %d\n", pid);
}
void do_parent_thing (void)
{
    printf ("I am the parent. My child's PID is %d\n", pid);
}

void main (void)
{
    switch (pid = fork())
    {
        case -1:
            printf ("fork failed\n");
            break;
        case 0:
            do_child_thing();
            break;
        default:
            do_parent_thing();
    }
    exit (0);
}
```

Listing 3.1
Trivial Example of Fork.

Since both processes are executing the same code, they both continue from the return from fork() (this is what's so bizarre!). In order to distinguish parent from child, fork() returns a function value of 0 to the child process but returns the PID of the child to the parent process. Listing 3.1 is a trivial example of the fork call.

The execve() Function

Of course, what really happens 99% of the time is that the child process invokes a new program by calling execve() to load an executable image file from disk. Listing 3.2 shows in skeletal form a simple command line interpreter. It reads a line of text from stdin, parses it, and calls fork() to create a new process. The child then calls execve() to load a file and execute the command just entered. execve() overwrites the calling process's code, data, and SSs.

If this is a normal "foreground" command, the command interpreter must wait until the command completes. This is accomplished with waitpid() which blocks the calling process

```
#include <unistd.h>

void main (void)
{
        char *argv[10], *filename;
        char text[80];
        char foreground;
        pid_t pid;
        int status;

        while (1)
        {
                gets (text);
// Parse the command line to derive filename and
// arguments. Decide if it's a foreground command.
                switch (pid = fork())
                {
                        case -1:
                                printf ("fork failed\n");
                                break;
                        case 0:  // child process
                                if (execve (filename, argv, NULL) < 0)
                                        printf ("command failed\n");
                                break;
                        default: // parent process
                                if (foreground)
                                        waitpid (&status, pid);
                }
        }
}
```

Listing 3.2
Command Line Interpreter.

until the process matching the `pid` argument has completed. Note by the way that most multitasking OSs do not have the ability to block one process or task pending the completion of another.

If `execve()` succeeds, it does not return. Instead, control is transferred to the newly loaded program.

The Linux File System

The Linux file system is in many ways similar to the file system you might find on a Windows PC or a Macintosh. It's a hierarchical system that lets you create any number of subdirectories under a *root directory* identified by "/". Like Windows, file names can be very long. However, in Linux, as in most Unix-like systems, filename "extensions"—the part of the filename following "."—have much less meaning than they do in Windows. For example, while Windows executables always have the extension ".exe" but Linux executables rarely have an extension at all. By and large, the contents of a file are identified by a file header rather than a specific extension identifier. Nevertheless, many applications, the C compiler for example, do support default file extensions.

Unlike Windows, file names in Linux are *case sensitive.* Foobar is a different file from foobar and is different from fooBar. Sorting is also case sensitive. File names beginning with upper case letters appear before those that begin with lower case letters in directory listings sorted by name.[3] File names that begin with "." are considered to be "hidden" and are not displayed in directory listings unless you specifically ask for them.

Additionally, the Linux filesystem has a number of features that go beyond what you find in a typical Windows system. Let's take a look at some of the features that may be of interest to embedded programmers.

File Permissions

Because Linux is multiuser, every file has a set of *permissions* associated with it to specify what various classes of users are allowed to do with that file. Get a detailed listing of some Linux directory, either by entering the command ls –l in a command shell window or with the desktop file manager. Part of the entry for each file is a set of 10 flags and a pair of names that look something like this:

```
-rw-r-r-   Andy   physics
```

In this example, Andy is the *owner* of the file and the file belongs to a *group* of users called physics, perhaps the physics department at some university. Generally, but not always, the owner is the person who created the file.

The first of the 10 flags identifies the file type. Ordinary files get a dash here. Directories are identified by "d", links are "l", and so on. We'll see other entries for this flag when we get to device drivers later. The remaining nine flags divide into three groups of three flags each. The flags are the same for all groups and represent respectively permission to read the file, "r", write the file, "w", or execute the file if it's an executable, "x". Write permission also allows the file to be deleted.

The three groups then represent the permissions granted to different classes of users. The first group identifies the permissions granted to the owner of the file and virtually always allows reading and writing. The second flag group gives permissions to other members of the same group of users. In this case, the physics group has read access to the file but not write access. The final flag group gives permissions to the "world," i.e., all users.

---

[3] The graphical file manager offers case insensitive sort as an option.

The "x" permission is worth a second look. In Windows, a file with the extension .exe is *assumed* to be executable. In Linux, a binary executable is identified by the "x" permission since we don't have an explicit file extension to identify it. Furthermore, only those classes of users whose "x" permission is set are allowed to invoke execution of the file. So if I'm logged in as an ordinary user, I'm not allowed to invoke programs that might change the state of the overall system such as changing the network address.

Another interesting feature of "x" is that it also applies to shell scripts, which we'll come to later in this chapter. For you DOS fans, a shell script is the same thing as a .bat file. It's a text file of commands to be executed as a program. But the shell won't execute the script unless its "x" bit is set.

The "root" User

There's one very special user, named "root," in every Linux system. Root can do anything to any file regardless of the permission flags. Root is primarily intended for system administration purposes and is not recommended for day-to-day use. Clearly, you can get into a lot of trouble if you're not careful, and root privileges pose a potential security threat. Nevertheless, the kinds of things that embedded and real-time developers do with the system often require write or executable access to files owned by root and thus require you to be logged in as the root user.

In the past, I would just log in as root most of the time because it was less hassle. One consequence of this is that every file I created was owned by root and couldn't be written by an ordinary user without changing the permissions. It became a vicious circle. The more I logged in as root, the more I *had* to log in as root to do anything useful. I've since adopted the more prudent practice of logging in as a normal user and only switching to root when necessary.

If you're logged on as a normal user, you can switch to being root with either the su, **substitute user**, or sudo commands. The su command with no arguments starts up a shell with root privileges provided you enter the correct root password. To return back to normal user status, terminate the shell by typing ^d or exit.

The sudo command allows you to execute a command as root provided you are properly authorized to do so in the "sudoers file," /etc/sudoers. The sudoers file is of course owned by root, so only the root user can authorize sudoers. Once you've been authenticated, by entering your own password, you may use sudo for a short period of time (default 5 min) without reentering your password.

For example, if I wanted to change the permissions on a file in the /dev directory, I could execute:

```
sudo chmod o + rw/dev/ttyS0
```

```
ls -l /proc
total 0
dr-xr-xr-x   3 root     root                0 Aug 25 15:23 1
dr-xr-xr-x   3 root     root                0 Aug 25 15:23 2
dr-xr-xr-x   3 root     root                0 Aug 25 15:23 3
dr-xr-xr-x   3 bin      root                0 Aug 25 15:23 303
dr-xr-xr-x   3 nobody   nobody              0 Aug 25 15:23 416
dr-xr-xr-x   3 daemon   daemon              0 Aug 25 15:23 434
dr-xr-xr-x   3 xfs      xfs                 0 Aug 25 15:23 636
dr-xr-xr-x   4 root     root                0 Aug 25 15:23 bus
-r--r--r--   1 root     root                0 Aug 25 15:23 cmdline
-r--r--r--   1 root     root                0 Aug 25 15:23 cpuinfo
-r--r--r--   1 root     root                0 Aug 25 15:23 devices
-r--r--r--   1 root     root                0 Aug 25 15:23 filesystems
dr-xr-xr-x   2 root     root                0 Aug 25 15:23 fs
dr-xr-xr-x   4 root     root                0 Aug 25 15:23 ide
-r--r--r--   1 root     root                0 Aug 25 15:23 interrupts
-r--r--r--   1 root     root                0 Aug 25 15:23 ioports
```

Figure 3.8
The /proc File System.

I would be prompted for my password and, if entered successfully, the command would be executed. I could then continue to execute sudo commands for 5 min without having to reenter my password. Note that sudo offers better security than su because the root user must authorize sudoers and the root password doesn't need to be disclosed.

The /proc File System

The /proc file system is an interesting feature of Linux. It acts just like an ordinary file system. You can list the files in the /proc directory, you can read and write the files, but they don't really exist. The information in a /proc file is generated on the fly when the file is read. The kernel module that registered a given /proc file contains the functions that generate read data and accept write data.

/proc files are a window into the kernel. They provide dynamic information about the state of the system in a way that is easily accessible to user-level tasks and the shell. In the abbreviated directory listing of Figure 3.8, the directories with number labels represent processes. Each process gets a directory under /proc with several directories and files describing the state of the process.

Try it Out

It's interesting to see how many processes Linux spawns just by booting up. Reboot your system, open a command shell and execute:

```
ps -A | wc
```

The ps command lists the processes running on the system. The output of ps is one line per process. The wc command counts the number of lines, words, and characters passed to it. The number of lines is essentially the number of processes running. On my Fedora 14 system, there are 281 processes running.

Now try:

```
ps -A | more
```

This command lets you see the output of ps one page at a time. Note that the information for the ps command comes from the /proc file system.

Here's another command that highlights the real-time nature of /proc data. Execute:

```
cat /proc/interrupts
```

The interrupts file lists all of the interrupt sources in the system, the number of times each interrupt has fired off since the system was booted, and the driver registered to the interrupt. Now, execute the command again:

```
cat /proc/interrupts
```

You'll see that many of the numbers have gone up, thus proving that the data is being generated dynamically.

The Filesystem Hierarchy Standard

A Linux system typically contains a very large number of files. For example, a typical Fedora installation may contain around 30,000 files occupying several gigabytes of disk space. Clearly, it's imperative that these files be organized in some consistent, coherent manner. That's the motivation behind the Filesystem Hierarchy Standard (FHS). The standard allows both users and software developers to "predict the location of installed files and directories."[4] FHS is by no means specific to Linux. It applies to Unix-like OSs in general.

The directory structure of a Linux file system always begins at the *root*, identified as "/". FHS specifies several directories and their contents directly subordinate to the root. This is illustrated in Figure 3.9. The FHS starts by characterizing files along two independent axes:

- *Sharable vs. non-sharable.* A networked system may be able to mount certain directories through Network File System (NFS) such that multiple users can share executables. However, some information is unique to a specific computer and is thus not sharable.

---

[4] *Filesystem Hierarchy Standard—Version 2.3 dated 1/29/04*, edited by Rusty Russell, Daniel Quinlan, and Christopher Yeoh. Available from www.pathname.com/fhs.

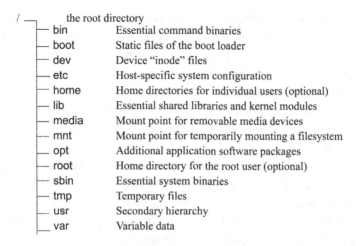

```
/ ——    the root directory
        — bin           Essential command binaries
        — boot          Static files of the boot loader
        — dev           Device "inode" files
        — etc           Host-specific system configuration
        — home          Home directories for individual users (optional)
        — lib           Essential shared libraries and kernel modules
        — media         Mount point for removable media devices
        — mnt           Mount point for temporarily mounting a filesystem
        — opt           Additional application software packages
        — root          Home directory for the root user (optional)
        — sbin          Essential system binaries
        — tmp           Temporary files
        — usr           Secondary hierarchy
        — var           Variable data
```

Figure 3.9
File System Hierarchy.

- *Static vs. variable.* Many of the files in a Linux system are executables that don't change, they're *static*. But the files that users create or acquire, by downloading or e-mail for example, are *variable*. These two classes of files should be cleanly separated.

Here is a description of the directories defined by FHS:

- /bin: Contains binary executables of commands used both by users and the system administrator. FHS specifies what files /bin must contain. These include among other things the command shell and basic file utilities. /bin files are static and sharable.
- /boot: Contains everything required for the boot process except configuration files and the map installer. In addition to the kernel executable image, /boot contains data that is used before the kernel begins executing user-mode programs. /boot files are static and non-sharable.
- /etc: Contains host-specific configuration files and directories. With the exception of mtab, which contains dynamic information about file systems, /etc files are static. FHS identifies three optional subdirectories of /etc:
    - /opt: Configuration files for add-on application packages contained in/opt.
    - /sgml: Configuration files for Standard Generalized Markup Language (SGML) and Extended Markup Language (XML).
    - /X11: Configuration files for X windows.
  In practice, most Linux distributions have many more subdirectories of /etc representing optional startup and configuration requirements.
- /home: (optional) Contains user home directories. Each user has a subdirectory under home with the same name as his/her user name. Although FHS calls this optional, in fact it is almost universal among Unix systems. The contents of subdirectories under /home is of course variable.

- /lib: Contains those shared library images needed to boot the system and run the commands in the root file system, i.e., the binaries in /bin and /sbin. In Linux systems /lib has a subdirectory, /modules, that contains kernel loadable modules.
- /media: Mount point for removable media. When a removable medium is auto-mounted, the mount point is usually the name of the volume.
- /mnt: Provides a convenient place to temporarily mount a file system.
- /opt: Contains optional add-in software packages. Each package has its own subdirectory under /opt.
- /root: Home directory for the root user.[5] This is not a requirement of FHS but is normally accepted practice and highly recommended.
- /sbin: Contains binaries of utilities essential for system administration such as booting, recovering, restoring, or repairing the system. These utilities are normally only used by the system administrator, and normal users should not need /sbin in their path.
- /tmp: Temporary files.
- /usr: Secondary hierarchy, see below.
- /var: Variable data. Includes spool directories and files, administrative and logging data, and transient and temporary files. Basically, system-wide data that changes during the course of operation. There are a number of subdirectories under /var.

The /usr Hierarchy

/usr is a secondary hierarchy that contains user-oriented files. Figure 3.10 shows the subdirectories under /usr. Several of these subdirectories mirror functionality at the root. Perhaps, the most interesting subdirectory of /usr is /src for source code. This is where the Linux source is generally installed. You may in fact have sources for several Linux kernels installed in /src under subdirectories with names of the form:

```
linux-<version number>-ext
```

You would then have a logical link named linux pointing to the kernel version you're currently working with.

"Mounting" File Systems

A major difference between Windows and Linux file systems has to do with how file-structured devices, hard disks, floppy drives, CD-ROMs, etc. are mapped into the system's directory or hierarchy structure. The Windows file system makes devices explicitly visible, identifying them with a letter-colon combination as in "C:". Linux however emphasizes a *unified* file system in which physical devices are effectively rendered invisible.

---

[5] Note the potential for confusion here. The directory hierarchy has both a *root*, "/", and a *root directory*, "/root", the home directory for the *root user*.

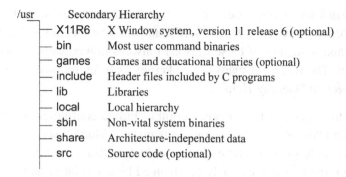

/usr   Secondary Hierarchy
        — X11R6   X Window system, version 11 release 6 (optional)
        — bin   Most user command binaries
        — games   Games and educational binaries (optional)
        — include Header files included by C programs
        — lib   Libraries
        — local   Local hierarchy
        — sbin   Non-vital system binaries
        — share   Architecture-independent data
        — src   Source code (optional)

Figure 3.10
/usr Hierarchy.

The mechanism that maps physical devices into the file system's directory structure is called "mounting."[6] Removable media devices such as the CD-ROM drive are the most visible manifestation of this feature. Before you can read a CD-ROM, you must mount the drive onto an existing node in the directory structure using the mount command as in:

```
mount/media/cdrom
```

This command works because the file /etc/fstab has information about what device is normally mounted at /media/cdrom and what type of file system, in this case iso9660, the device contains.

Like file permissions, mount can sometimes be a nuisance if all you want to do is read a couple of files off a CD. But the real value of the mount paradigm is that it isn't limited to physical devices directly attached to the computer nor does it only understand native Linux file systems. As we'll see later, we can mount parts of file systems on remote computers attached to a network to make their files accessible on the local machine. It's also possible to mount a device that contains a DOS FAT or VFAT file system. This is particularly useful if you build a "dual-boot" system that can boot up into either Windows or Linux. Files can be easily shared between the two systems.

System Configuration

The section above on the FHS mentioned the /etc directory. Here's one place where Unix systems really shine relative to Windows. Okay, there may be any number of ways that

---

[6] The term probably harks back to the days of reel-to-reel tape drives onto which a reel of tape had to be physically "mounted."

Unix outshines Windows. In any case, the /etc directory contains essentially all of the configuration information required by the kernel and applications in the form of plain text files. The syntax and semantics of these files isn't always immediately obvious, but at least you can read them. The format of many of the kernel's configuration files is documented in man pages (see Section "Getting Help").

By contrast, Windows systems have made a point of hiding configuration information in a magical place called "the Registry" (cue ominous music). Mortal users are often cautioned to stay away from the Registry because improper changes can render the system unbootable. In fact, the Registry can only be changed by a special program, regedit, and it was only about a year ago that I finally figured out where the Registry physically resides.

If you've ever mustered up the courage to run regedit and actually look at the Registry, you've no doubt noticed that the entries are rather obscure symbols and the values are even more obscure binary values. I submit that this was absolutely intentional on Microsoft's part. They didn't want idiot consumers mucking around with their computers and possibly bricking[7] them. Linux users, however, know what they're doing (it says here) and are entitled to see and modify configuration data to their heart's content.

The Shell

One of the last things that happens as a Linux system boots up is to invoke the command interpreter program known as the *shell*. Its primary job is to parse commands you enter at the console and execute the corresponding program. But the shell is much more than just a simple command interpreter. It incorporates a powerful, expressive interpretive programming language of its own. Through a combination of the shell script language and existing utility programs it is quite possible to produce very sophisticated applications without ever writing a line of C code. In fact, this is the general philosophy of Unix programming. Start with a set of simple utility programs and link them together through the shell scripting language.

The shell scripting language contains the usual set of programming constructs for looping, testing, functions, and so on. But perhaps the neatest trick in the shell bag is the concept of "pipes." This is a mechanism for streaming data from one program to another. The metaphor is perfect. One program dumps bytes into one end of the pipe while a second program takes the bytes out the other end.

---

[7] "Bricking" refers to messing up the configuration of a computer so badly that it no longer boots, thus rendering it the equivalent of a brick. The term is used alot in the embedded space. A more apt term for desktop boxes might be "boat anchoring."

Most Linux/Unix utility programs accept their input from a default device called "stdin". Likewise, they write output to a device called "stdout". Any error messages are written to "stderr". Normally, stdin is the keyboard while stdout and stderr are the display. But we can just as easily think of stdin and stdout as two ends of a pipe.

stdin and stdout can be *redirected* so that we can send output to, for example, a file or take input from a file. In a shell window, try typing just the command cat with no arguments. cat[8] is perhaps the simplest of all Unix utilities. All it does is copy stdin to stdout, line by line. When you enter the command with no arguments, it patiently waits at stdin for keyboard input. Enter a line of text and it will send that line to stdout, the display. Type Ctrl-C to exit the program.

Now try this:

```
cat>textfile
```

Enter a line of text. This time you don't see the line repeated because the " > " operator has *redirected* stdout to the file textfile. Type a few lines, then Ctrl-C to exit.

The final step in this exercise is:

```
cat<textfile
```

Voila! The file you created with the previous command shows up on the screen because the " < " operator redirected stdin to textfile. cat actually implements a shortcut so that if you enter a filename as a command line argument, without the < operator, it takes that as an input file. That is:

```
cat textfile
```

The real power of pipes though is the "|" operator, which takes stdout of one program and feeds it to stdin of another program. When I did the above exercises, I created a textfile containing

```
this is a file
of text from the keyboard
```

Now if I execute

```
cat textfile | grep this
```

I get:

```
this is a file
```

---

[8] Short for concatenate. Don't you just love Unix command names?

grep, as you may have guessed, is a filter. It attempts to match its command line arguments against the input stream stdin. Lines containing the argument text are passed to stdout. Other lines are simply ignored. What happened here is that the output of cat became the input to grep.

In typical Unix fashion, grep stands for Get Regular Expression "something." I forget what the "p" stands for. Regular expressions are in fact a powerful mechanism, a scripting language if you will, for searching text. grep embodies the syntax of regular expressions.

Shell scripts and makefiles make extensive use of redirection and piping.

Some other shell features are worth a brief mention because they can save a lot of typing. The shell maintains a command history that you can access with the up arrow key. This allows you to repeat any previous command, or edit it slightly to create a new, similar command. There is a history command that dumps out the accumulated command history of the system or some subset. The history is maintained in the file. bash_history in your home directory.

Figure 3.11 shows the last few lines of the history command on my system. To reexecute any command in the list, simply enter "!" followed by the command number in the history list as in:

    !998

to reexecute history.

Another really cool feature is auto-completion, which attempts to complete filename arguments for you. Say I wanted to execute

    cat verylongtextfilename

I could try entering

    cat verylong<tab>

Provided the remainder of the filename is unique in the current directory, the shell will automatically complete it, saving me the trouble of typing the whole filename. The shell beeps if it can't find a unique match. Then you just have to type a little more until the remainder is unique.

Finally, the "~" character represents your home directory. So from anywhere in the file system you can execute cd~ to go to your home directory.

There are in fact several shell programs in common use. They all serve the same basic purpose yet differ in details of syntax and features. The most popular are as follows:

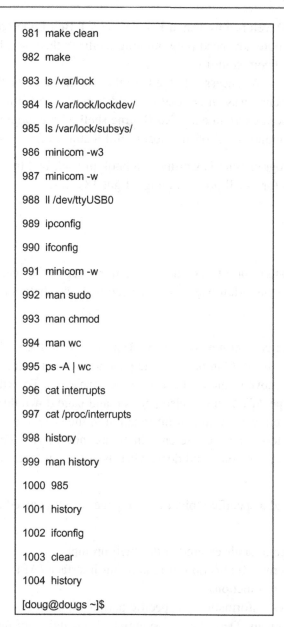

Figure 3.11
`history` Command.

• Bourne Again Shell—`bash`. `bash` is a "reincarnation" of the Bourne shell, `sh`, originally written by Stephen Bourne and distributed with Unix version 7. It's the default on most Linux distributions and you should probably use it unless you have good reason, or strong feelings, for using a different shell.

- Korn Shell—ksh. Developed by David Korn at AT&T Bell Laboratories in the early 1980s. It features more advanced programming facilities than bash but nevertheless maintains backward compatibility.
- Tenex C Shell—tcsh. A successor to the C shell, csh, that was itself a predecessor to the Bourne shell. Tenex was an OS that inspired some of the features of tcsh.
- Z Shell —zsh. Described as an extended Bourne shell with a large number of improvements, including some of the most useful features of bash, ksh, and tcsh.

The subject of shell programming is worthy of a book in itself and there are many. When I searched Amazon.com for "shell programming," I got 189 hits.

Getting Help

The official documentation for a Unix/Linux system is a set of files called "man pages," man being short for manual. Man pages are accessed with the shell command man as in:

```
man cp
```

To get help on the shell copy command. Try it. Type man man at the shell prompt to learn more about the man command. Man presents the page a screen at a time with a ":" prompt on the bottom line. To move to the next screen, type <space>. To exit man before reaching the end of the page, type "q". You can also page up and page down through the man page. Interestingly enough, you won't find that information in the man man page. The writing style in man pages is rather terse; they're reference in nature, not tutorial. The information is typically limited to: what the command does, what its arguments are, and what options it takes.

To make it easier to find a specific topic, the man pages are organized into sections as follows:

Section 1: User commands entered at the shell prompt.
Section 2: The kernel Application Programming Interface (API) functions.
Section 3: C library functions.
Section 4: Devices. Information on specific peripheral devices.
Section 5: File formats. Describes the syntax and semantics for many of the files in /etc.
Section 6: Games.
Section 7: Miscellaneous.
Section 8: System administration. Shell commands primarily used by the system administrator.

Another useful source of information is "info pages." Info pages tend to be more verbose, providing detailed information on a topic. Info pages are accessed with the info command. Try this one:

```
info gcc
```

to learn more about the GCC compiler package.

In the graphical desktop environments, KDE and GNOME, you can also access the man and info pages graphically. I find this especially useful with the info pages to find out what's there.

Finally, no discussion of getting help for Linux would be complete without mentioning Google. When you're puzzling over some strange behavior in Linux, Google is your friend. One of my common frustrations is error messages because they rarely give you any insight into what really went wrong. So type a part of the error message into the Google search box. You'll likely get back at least a dozen hits of forum posts that deal with the error and chances are something there will be useful.

With a good basic understanding of Linux, our next task is to configure the development workstation and install the software that will allow us to develop target applications.

Resources

Sobell, Mark G., *A Practical Guide to Linux*. This book has been my bible and constant companion since I started climbing that steep Linux learning curve. It's an excellent beginner's guide to Linux and Unix-like systems in general, although having been published in 1997 it is getting a bit dated and hard to find. It has been superseded by Sobell, Mark G., *A Practical Guide to Linux Commands, Editors, and Shell Programming*.

tldp.org—The Linux Documentation Project. As the name implies, this is the source for documentation on Linux. You'll find how-to's, in-depth guides, FAQs, man pages, even an online magazine, the *Linux Gazette*. The material is generally well written and useful.

## CHAPTER 4

The Host Development Environment

*When you say "I wrote a program that crashed Windows," people just stare at you blankly and say "Hey, I got those with the system, **for free**."*

Linus Torvalds

In this chapter, we'll configure our host workstation environment in preparation for configuring the target board in Chapter 5. The steps involved include:

- Installing some software from the Mini2440 DVD
- Configuring the workstation
- Configuring networking on the workstation

We've got a lot to do in this chapter. But before we dive into that, let's take a look at the cross-development environment in general.

Cross-Development Tools—The GNU Tool Chain

Not only is the target computer limited in resources, it may be a totally different processor architecture from your (probably) x86-based development workstation. Thus, we need a cross-development tool chain that runs on the PC but may have to generate code for a different processor. We also need a tool that will help us debug code running on that (possibly different) target.

GCC

By far the most widely used compiler in the Linux world is the GNU compiler collection (GCC). It was originally called the GNU C compiler but the name was changed to reflect its support for more than just the C language. GCC has language front ends for C, C++, Objective C, Fortran, Java, and Ada as well as run-time libraries for these languages.

GCC also supports a wide range of target architectures in addition to the x86. Supported targets include:

- Alpha
- ARM
- M68000
- MIPS

- PowerPC
- SPARC

among others. In fact, the Wikipedia article on GCC lists 43 processors and processor families supported by the FSF version of GCC.

GCC can run in a number of operating environments including Linux and other variants of Unix. There are even versions that run under DOS and Windows.

GCC can be built to run on one architecture (a PC for example) while generating code for a different architecture (an ARM perhaps). This is the essence of cross development.

GCC is an amalgam of compiler, assembler, and linker. Individual language front ends translate source code into an intermediate assembly language format that is then assembled by a processor-specific assembler and ultimately linked into an executable. For a program consisting of a single C source file, the command to build an executable can be as simple as:

```
gcc myprogram.c
```

By default, this creates an executable called a.out. The −o option lets you do the obvious thing and rename the executable thus:

```
gcc −o myprogram myprogram.c
```

If your project consists of multiple source files, the −c option tells GCC to compile only.

```
gcc −c myprogram.c
```

generates the relocatable object file myprogram.o. Later this can be linked with other .o files with another GCC command to produce an executable.

Generally though, you won't be invoking GCC directly. You'll use a makefile instead.

Make

Real-world projects consist of anywhere from dozens to thousands of source files. These files have dependencies such that if you change one file, a header for example, another file, a C source, needs to be recompiled and ultimately the whole project needs to be rebuilt. Keeping track of these dependencies and determining what files need to be rebuilt at any given time is the job of a powerful tool called the Make utility.

In response to the shell command make, the Make utility takes its instructions from a file named Makefile or makefile in the current directory. This file consists of *dependency lines* that specify how a *target file* depends on one or more *prerequisite files*. If any of the prerequisite files are more recent than the target, make updates the target using one or more *construction commands* that follow the dependency line. The basic syntax is:

```
target: prerequisite-list
<tab> construction commands
```

One common mistake to watch out for when writing makefiles is substituting spaces for the leading tab in the construction commands. It *must* be a tab.

Don't let this simple example fool you. There is a great deal of complexity and subtlety in the make syntax, which looks something like the shell scripting language. The sample code that you'll download later in the chapter has some simple examples of makefiles. In a later chapter, you'll configure and build the Linux kernel, which has a huge, very involved makefile.

Often the best way to create a makefile for a new project is to start with the one for an existing, similar project.

GDB

GDB stands for the "GNU DeBugger." This is a powerful source-level debugging package that lets you see what's going on inside your program. You can step through the code, set breakpoints, examine and change variables, and so on. Like most Linux tools, GDB itself is command-line driven, making it rather tedious to use. There are several graphical front ends for GDB that translate graphical user interface (GUI) commands into GDB's text commands. Eclipse, which you'll encounter in Chapter 6, has an excellent graphical front end for GDB.

GDB can be set up to run on a host workstation while debugging code on a separate target machine. The two machines can be connected via serial or network ports, or you can use an in-circuit emulator (ICE). Many ICE vendors have GDB back ends for their products.

There are also specialized debug ports built into some processor chips. These go by the names Joint Test Action Group (JTAG) and Background Debug Mode (BDM). These ports provide access to debug features such as hardware breakpoint registers. GDB back ends exist for these ports as well.

Install Software

What's on the DVD?

The recommended target board for the examples in the book is the Mini2440 produced by FriendlyARM and distributed in the United States by Industrial ARMworks. The resources section has links for ordering. Chapter 5 contains more details on the board.

The DVD that comes with the Mini2440 contains a veritable plethora of software. In addition to Linux, it has support for Windows CE and MicroC/OS II. Our primary interest is the `linux/` subdirectory that includes the following:

- `arm-linux-gcc-4.3.2.tgz`—A GCC cross-tool chain for building Mini2440 software.
- `arm-qtopia.tgz`—Application development kit for the QT graphics library. This one builds code for the Mini2440.
- `bootloader.tgz`—Sources for both u-boot and the "vivi" boot loader contained in the NOR flash. There is a later version of u-boot that we'll explore in a later chapter.
- `busybox-1.13.3-mini2440.tgz`—Source code for the BusyBox utility that we'll look at in a later chapter.
- `Descriptions.txt`—What you would expect. Unfortunately, it's in Chinese.
- `examples.tgz`—Sample code that exercises many of the peripherals and ports on the board.
- `linux-2.6.29-mini2440-20090708.tgz`—A fork of the Linux kernel pre-patched for the Mini2440. We'll use a later kernel that incorporates the Mini2440.
- `logomaker.tgz`—An executable. I'm guessing it creates image files for the splash screen.
- `madplay-0.15.2b.tar.gz`—An MPEG audio decoder and player. Complete source code.
- `mkyaffs2image.tgz`—Two executables for creating YAFFS images. You'll want to use the one named mkyaffs2image—128M.
- `root_qtopia.tgz`—A root file system for the Mini2440. We'll mount it over NFS.
- `vboot-src-20090721.tgz`—Appears to be a boot loader of some kind. Doesn't include any documentation.
- `wireless_tools.29.tar.gz`—Tools to manipulate the wireless extensions, an interface for setting wireless LAN-specific parameters.
- `x86-qtopia.tgz`—Another QT application development package. This one generates code for the x86.

Note that the above list is the contents of the `linux/` directory as of December 2011. It could change with later releases.

Install Cross-Tool Chain

Mount the DVD.[1] In KDE, simply inserting the DVD brings up a dialog called "Available Devices." Click on the entry `UDF Volume`. You're offered three "actions," the third being `Open with File Manager`. Select that one. This effectively mounts the DVD at `/media/UDF Volume/` and opens that location in a file manager window, which we won't really use.

---

[1] If you're running a virtual machine, be sure the VM is attached to the optical drive.

In a shell window, become root user with either the `su` or `sudo` commands and execute the following set of commands:

```
cd /
cp /media/"UDF Volume"²/arm-linux-gcc-4.3.2.tgz.
tar —xzf arm-linux-gcc-4.3.2.tgz
rm arm-linux-gcc-4.3.2.tgz
```

`cp /media/"UDF Volume"²/arm-linux-gcc-4.3.2.tgz.` (Note the dot that represents the current directory.)

`tar —xzf arm-linux-gcc-4.3.2.tgz` (This will take a while. It's a *big* file.)

Don't forget tab auto-completion to save yourself a lot of typing. The `tar` command uncompresses a compressed Tape ARchive file.

Now if you go to `/usr/local/`, you'll see a subdirectory, `arm/`, that wasn't there before. Under that is subdirectory `4.3.2/` that contains a GCC cross-tool chain. This is a "lite" version, i.e. free, of CodeBench, a commercial product from CodeSourcery. It is one of the most popular cross-tool chains for the ARM architecture.

You'll need to add `/usr/local/arm/4.3.2/bin` to your PATH, the list of directories that the shell searches to find executables. In Fedora, the file that defines the PATH is `.bash_profile`. In Ubuntu/Debian installations, it's just `.profile`. From here on we'll be doing a lot of editing. If you have a favorite Linux editor, by all means use it. Personally, I refuse to touch vi or emacs.[3] My favorite KDE editor is Kwrite. Double-click a text file in a file manager window and it automatically opens in Kwrite.

In `.bash_profile` (or `.profile`) you'll find a line that begins `PATH = $PATH:` probably followed by `$HOME/bin`. At the end of that line add "`:/usr/local/arm/4.3.2/bin`". Save the file and close the editor. The new path won't be in effect until you log out and log back in again. No hurry. It will be a while yet before we get around to building a program.

Install Root File System

The other component that we need to install at this point is the root file system that the target board will mount over NFS. You should still be logged in as root. Execute the following command sequence:

```
cd /home/<your_home_directory>
cp /media/UDF\ Volume/root_qtopia.tgz.
tar —xzf root_qtopia.tgz
rm root_qtopia.tgz
```

---

[2] Linux utilities don't get along well with spaces in file or directory names. Use quotes to bracket a name with spaces or precede the space with a "\" character.

[3] Although I have had my arm twisted to try vim and it's not too bad.

Note that you can't use "~" here to specify the home directory because you're logged in as root and "~" evaluates to /root. Take a look at the new root_qtopia/ subdirectory in your home directory and you'll see it's virtually identical in layout to the root directory of your workstation.

The Terminal Emulator, minicom

minicom is a fairly simple Linux application that emulates a dumb RS-232 terminal through a serial port. This is what we'll use to communicate with the target board.

Recent Linux distributions tend not to install minicom by default. It's easy enough to install. As root user (again, either su or sudo) execute:

```
yum install minicom       in Fedora
apt-get install minicom   in Ubuntu or Debian
```

Both of these commands do essentially the same thing. They go out on the Internet to find the specified package, resolve any dependencies, download the packages, and install them.

Once minicom is installed, there are a number of configuration options that we need to change to facilitate communication with the target.

In a shell window, again *as root user*, enter the command minicom-s. If you're running minicom for the first time you may see the following warning message:

```
WARNING: Configuration file not found. Using defaults
```

You will be presented with a configuration menu. Select serial port setup (Figure 4.1). By default, minicom communicates through the modem device, /dev/modem. We need to change that to talk directly to one of the PC's serial ports Type "A" and replace the word "modem" with either "ttyS0" or "ttyS1," where ttyS0 represents serial port COM1 and ttyS1 represents COM2. However, if your host only has USB ports and you're using a USB to serial converter, the correct device is most likely "ttyUSB0." Again, if you're running a virtual machine, be sure the VM is attached to the serial port.

Type "E" followed by "I" to select 115,200 baud. Make sure both hardware and software flow control are set to "no." These are toggles. Typing "F" or "G" will toggle the corresponding value between "yes" and "no." Figure 4.1 shows the final serial port configuration.

Type <Enter> to exit serial port setup and then select modem and dialing. Here we want to delete the modem's Init string and Reset string since they're just in the way on a direct serial connection. Type "A" and backspace through the entire Init string. Type "B" and do the same to the Reset string. The other strings in this menu don't matter because they'll never be invoked.

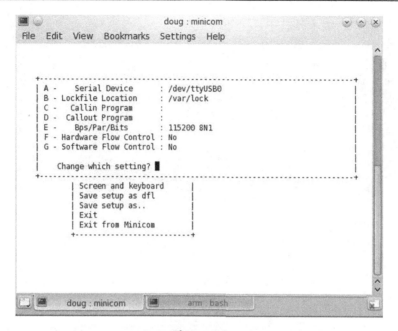

Figure 4.1
minicom Serial Port Setup Menu.

Type <Enter> to exit modem and dialing and then select screen and keyboard. Type "B" once to change the backspace key from BS to DEL.

Type <Enter> to exit screen and keyboard. Select Save setup as dfl (default). Then select Exit.

Execute ls-l /dev/ttyS0 (or /dev/ttyS1 or /dev/ttyUSB0 if that's the one you chose in serial port setup). Note that only the owner and the group have access to the device. In order to gain access to this port, you must become a member of the dialout group.

As is often the case, there are two ways of doing this and which you choose is a matter of personal preference. There's a graphical dialog accessible from the Fedora Start menu under Administration>Users and Groups. You'll be asked to enter the root password. That brings up the User Manager dialog shown in Figure 4.2. Highlight your user name and select Properties. Then select the Groups tab (Figure 4.3). Scroll down until you find dialout and check it. Click OK, then close the User Manager dialog.

The other approach is to manually edit the file /etc/group. As root user, open /etc/group in an editor. Find the line that begins "dialout:" and add ":<your_user_name>" at the end of the line. Save the file and close the editor.

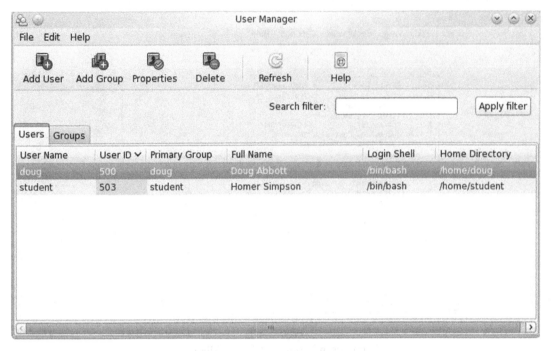

Figure 4.2
User Manager Dialog.

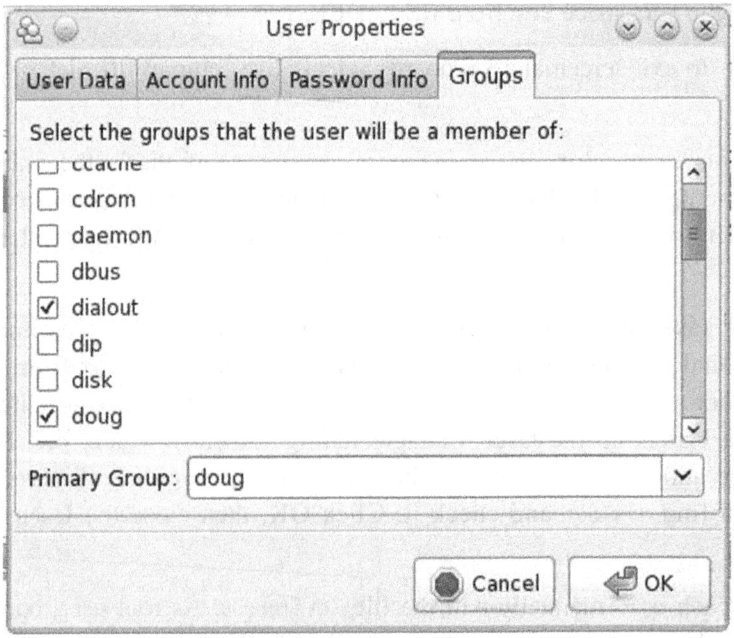

Figure 4.3
Groups Dialog.

Networking

There are a couple of changes needed on the host's network configuration.

Network Address

Your workstation is probably configured to get a network address via dynamic host configuration protocol (DHCP). But in this case, to keep things simple, we're going to specify fixed IP addresses for both the workstation and the target.

The Fedora graphical menu for changing network parameters is accessed from `Administration -> Network Configuration`.[4] In the Network Configuration dialog box, select the Devices tab (it's the default), highlight `eth0`, and click Edit. In the Ethernet Device dialog, General tab, unselect the "Automatically obtain IP address settings with …" box. Now enter the fixed IP address. Network address 192.168.1 is a good choice here because it's one of a range of network addresses reserved for local networks. Select node 2 for your workstation. Enter the Subnet Mask as shown in Figure 4.4. You may also want to set the Default Gateway Address and DNS nodes if you're connected to a network.

The checkbox Controlled by NetworkManager can be a problem. If your network port is already active and this box is checked, it's probably not an issue. But note when you exit back to the Network Configuration dialog, the Activate and Deactivate buttons are not "active." If the network port is not active, you have no control over it and it's unlikely to be auto-activated at boot time. What I have found to work through experimentation is to uncheck the NetworkManager box in the edit dialog, activate the port, then check the box again. The next time you boot, the port should auto-activate.

When you exit from the Network Configuration dialog, you're asked if you want to save the changes. Yes you do. You're also warned that you may have to restart the network. Click OK. My experience has been that you usually don't have to restart the network.

Alternatively, you can just go in and directly edit the network device parameters file. Network configuration parameters are found in `/etc/sysconfig/network-scripts/` where you should find a file named something like `ifcfg-eth0` that contains the parameters for network adapter 0. You might want to make a copy of this file and name it something like `dhcp-ifcfg-eth0`. That way you'll have a DHCP configuration file for future use if needed. Now open the original file with an editor (as root user of course). It should look something like Listing 4.1a. The lines shown here may be in a different order and interspersed with other lines.

---

[4] Configuration options have an annoying tendency to move around with each new release of Fedora. This is where it's located in Fedora 14.

Figure 4.4
Edit Network Device Dialog.

Change the line BOOTPROTO = dhcp to BOOTPROTO = none and add the four new lines as shown in Listing 4.1b. Strictly speaking, the Gateway entry is only necessary if the workstation is connected to a network with Internet access.

Ubuntu users will find that the graphical network configuration dialog, if indeed it exists at all, is virtually impossible to access. So you'll need to edit the appropriate configuration file, which is /etc/network/interfaces. Initially, this file has two lines describing the local loopback interface as shown in Listing 4.2a. Add the four new lines as shown in Listing 4.2b that describe eth0. Again, the Gateway entry is only necessary if the workstation is connected to a network with Internet access.

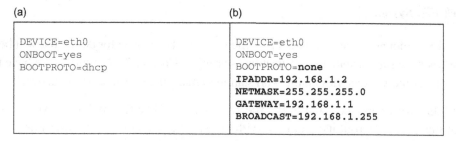

(a)

```
DEVICE=eth0
ONBOOT=yes
BOOTPROTO=dhcp
```

(b)

```
DEVICE=eth0
ONBOOT=yes
BOOTPROTO=none
IPADDR=192.168.1.2
NETMASK=255.255.255.0
GATEWAY=192.168.1.1
BROADCAST=192.168.1.255
```

Listing 4.1
(a) ifcfg-eth0 and (b) Revised ifcfg-eth0.

(a)

```
auto lo
iface lo inet loopback
```

(b)

```
auto lo
iface lo inet loopback

auto eth0
iface eth0 inet static
address 192.168.1.2
netmask 255.255.255.0
gateway 192.168.1.1
```

Listing 4.2
(a) Interfaces and (b) Revised Interfaces.

What About Wireless?

Personally, I have not had any success configuring wireless network ports directly in Linux. But wireless ports do work fine in a virtual machine environment because the VM manager virtualizes the network interface so that it looks like something else to the guest machine. VirtualBox for example translates any physical network interface to an Intel 82540EM gigabit interface.

Having two interfaces can be very useful. You could, for example, leave the wireless port attached to your network getting its IP address through DHCP and reconfigure the Ethernet port as described above for exclusive connection to the target. For example, my network is 192.168.1. I plug the target board into a switch. The DHCP server in the router serves node addresses ranging from 100 to 199. The Windows host OS on my laptop gets its IP address from DHCP. The Linux guest has its address fixed by the procedure described here and everything gets along fine.

Network File System

We use NFS to mount the target board's root file system.[5] The reason for doing that will become apparent when we start developing applications starting in Chapter 7. That means we have to "export" one or more directories on the workstation to make them visible on the network.

Exported directories are specified in the file /etc/exports. The file probably exists, but is empty at this point. As root user, open it with an editor and add the following line:

```
/home/<your_home_name> *(rw,no_root_squash,sync,no_subtree_check)
```

Replace <your_home_name> with the name of your home directory. This makes your home directory visible on the network where other nodes can mount parts of it to their local file system. The "*" represents clients that are allowed to mount this directory. In this case, the directory is visible to any machine on the network. You can set specific machines either by DNS name or IP address (as in machine1.intellimetrix.us, or 192.168.1.50, for example). You can also specify all the machines on a given network by giving the network and network mask as in 192.168.1.0/255.255.255.0.

The options inside the parentheses describe how the client machines can access the directory. Options shown here are:

- rw: Allow read and write access. Alternately, you can specify ro for read only.
- no_root_squash: By default, a file request made by the root user on a client machine is treated on the server as if it came from user nobody. If no_root_squash is specified, the root user on the client has the same level of access as the server's root user. This can have security implications, but is necessary in some circumstances.
- sync: Normally, file writing is asynchronous, meaning that the client will be informed that a write is complete when NFS hands the request over to the file system but before it actually reaches its final destination. The sync option waits until the operation really is complete.
- no_subtree_check: If only part of a volume is exported, a routine called *subtree checking* verifies that a requested file is in the exported part. If the entire volume is exported, or you're not concerned about this check, disabling subtree checking will speed up transfers.

NFS is somewhat picky about the format of this file. The last line must end with an end-of-line character and no spaces are allowed inside the parentheses.

With the exports file in place, we need to be sure that NFS is actually running. The Fedora graphical dialog for this is at Administration>Services (Service Management) as shown in Figure 4.5. Scroll down to nfs and select it. Initially nfs is probably disabled, so you'll need

---

[5] Ubuntu distros tend to not have NFS installed by default. Execute sudo apt-get install nfs-kernel-server to install it.

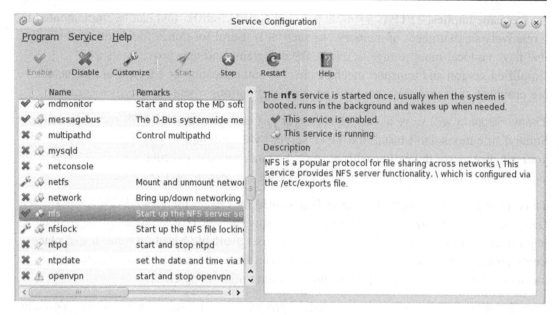

Figure 4.5
Service Management Dialog.

to enable and start it. If it is running, you should restart it to force it to reread the modified exports file.

From the shell, you can control nfs and other services with the `service` command, which runs a specified System V init script located in `/etc/init.d`. So, for example, to start nfs you would execute:

```
service nfs start
```

To stop it, you would execute:

```
service nfs stop
```

Check the status with:

```
service nfs status
```

Look in `/etc/init.d` to see that there is a script file named `nfs`. Have a look at the file to see how it works.

Trivial File Transfer Protocol

The last item of business for this chapter is installing (if necessary) and enabling the trivial file transfer protocol (TFTP). We'll use that in Chapter 5 for downloading some images to the target board's boot loader and later when building a new Linux kernel.

As the name implies, TFTP is a simple protocol for file transfer that can be implemented in a relatively small amount of memory. As such, it is useful for things like booting devices that have no local mass storage. It uses UDP datagrams and thus provides its own simplified session and transport mechanisms. No authentication or encryption mechanisms are provided, so it probably shouldn't be used in situations where security is a concern.

The workstation will serve as our TFTP server just as it provides the NFS server function. Some distributions don't install the TFTP server by default and so you must install it yourself. Just as we did earlier with `minicom`, Fedora users can execute `yum install tftp-server` while Ubuntu and Debian users can execute `apt-get install tftp-server`.

TFTP is one of several network protocols managed by `xinetd`, the eXtended InterNET Daemon. `xinetd` features access control mechanisms such as TCP Wrapper ACLs, extensive logging capabilities, and the ability to make services available based on time. It can place limits on the number of servers that the system can start, and has deployable defense mechanisms to protect against port scanners, among other things.

Once installed, TFTP needs to be enabled. There are two ways to accomplish that. You can fire up the Service Configuration dialog like we did for NFS. Scroll down to `tftp` and enable it. Then scroll down and make sure `xinetd` is running. The other approach is to edit the file `/etc/xinetd/tftp`. You'll need to be root user to save any changes. Find the line that starts "disable =" and change "yes" to "no."

While you're in this file, also note the line that starts "server_args =". The argument for this line is the default directory for TFTP transfers. Currently, in Fedora it is `/var/lib/tftpboot`. Any files you want to download to the target board over TFTP need to be copied to this directory. But this directory is owned by root and not writable by the world. So you need to make it writable by everyone by executing, as root user:

```
chmod 777 /var/lib/tftpboot
```

We now have almost everything we need on the host workstation for embedded development. In Chapter 5, we'll focus on configuring the target embedded board that we'll use for application development. Following that we'll install an integrated development environment (IDE) that will make application development a lot more productive.

Resources

andahammer.com—The web site for Industrial ARMworks, the US distributor of the Mini2440.
friendlyarm.net—Chinese manufacturer of the Mini2440. There is an extensive download page, but much of the useful stuff is already on the DVD. The sales page has a complete list of distributors worldwide.
Mecklenburg, Robert, *Managing Project with GNU Make, third Ed.*, O'Reilly, 2005. This is the bible when you're ready to fully understand and exploit the power of `make`.

## CHAPTER 5

The Hardware

Beware of programmers carrying soldering irons.

Anonymous

Embedded Hardware

Embedded computing covers an extremely wide range of hardware: from the (probably) 8-bit processor that powers your automatic coffee maker to the (perhaps multicore) 32-bit processor in your cell phone or tablet, up to the massively parallel processing environment that runs the telecommunications network. In this chapter, we'll get to work on the embedded target board that we'll use in the next few chapters for applications programming.

ARM Single Board Computer

ARM is a 32-bit reduced instruction set computer (RISC) architecture developed by ARM Holdings, a British company originally known as Advanced RISC Machines. It is said to be the most widely deployed 32-bit architecture in terms of numbers produced. The relative simplicity of ARM processors makes them suitable for low-power applications. As a result, they have come to dominate the mobile and consumer electronics market. About 98% of all mobile phones use at least one ARM processor.[1]

ARM Holdings itself does not make chips. Rather, it licenses the architecture to a number of semiconductor manufacturers including AppliedMicro, Atmel, Freescale, Marvell (formerly Intel), Nvidia, ST-Ericsson, Samsung, and Texas Instruments (TI), among others.

The next few chapters will explore application programming in a cross-platform environment using a single board computer (SBC) as the target. Specifically, we'll be using a small, ARM-based SBC, the Mini2440, based on the Samsung S3C2440 ARM9 system-on-a chip (SOC). The Mini35 version of the board includes a 3.5 in. touch screen LCD and comes with all necessary cables and a power supply, plus a DVD of useful software. The Mini35 is priced at around $120 in single quantities (Figure 5.1).

---

[1] c|net, "ARMed for the Living Room," http://news.cnet.com/ARMed-for-the-living-room/2100-1006_3-6056729.html.

Linux for Embedded and Real-time Applications.
© 2013 Elsevier Inc. All rights reserved.

DOI: http://dx.doi.org/10.1016/B978-0-12-415996-9.00005-8

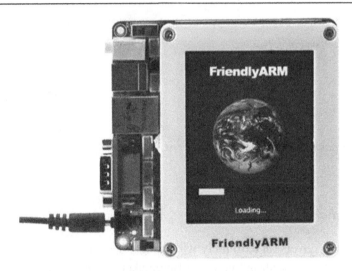

Figure 5.1
Mini2440 with 3.5 in. LCD.

Why the Mini2440? In researching ARM-based SBCs, I felt that the Mini2440 had the best value for the money. A diverse feature set at a good price.

The Mini2440 is made by a Chinese company named FriendlyARM. The US distributor is Industrial ARMworks. See the resources section for details.

Specifications

- Processor: Samsung S3C2440, 405 MHz
- 64 MB RAM
- 128 MB to 1 GB NAND flash
- 2 MB NOR flash
- Peripheral connections:
  - 10/100 Ethernet
  - USB 2.0 host and device
  - 3 RS-232 (1 9-pin D connector)
  - SD/MMC slot
  - Multichannel ADC (pot attached to channel 0)
  - Stereo audio out, mic in
  - JTAG
  - 4 user LEDs
  - 6 user pushbuttons
  - Power (5 VDC, barrel connector with switch, and LED)
- Size: 100 mm (~3.6 in.) × 100 mm

What About Other Boards?

There is a wide range of ARM-based SBCs on the market. I chose the Mini2440 because it seemed to be the best value. Nevertheless, it's worth reviewing a couple of popular alternatives.

BeagleBoard

The BeagleBoard is a popular Open Source hardware SBC based on the TI OMAP3530 SOC that combines an ARM Cortex-A8 CPU with a TI TMS320C64x digital signal processor. It was initially developed by TI as an educational tool, and the design is released under the Creative Commons license, making it effectively Open Source.

Specifications (Rev. C4)

- Processor: TI OMAP3530 at 720 MHz
- 256 MB RAM
- 256 MB NAND flash
- Peripheral connections:
  - DVI-D using HDMI connector, maximum resolution 1280×1024
  - S-video
  - USB 2.0 HS OTG
  - RS-232
  - JTAG
  - SD/MMC slot
  - Stereo audio in and out
  - Power (5 VDC, barrel connector)
- Size: 75 mm ($\sim$3 in.) $\times$ 75 mm
- Suggested distributor price: $125

Requiring only 2 W at 5 V, the board can also be powered through its USB port.

A modified version of the BeagleBoard called the BeagleBoard-xM started shipping in August 2010. Slightly larger at 8.25×8.25 cm^2, the -xM features a faster CPU core (1 GHz compared to 720 MHz for the C4), more RAM (512 MB vs. 256 MB), onboard Ethernet jack, camera port, and 4-port USB hub. The -xM lacks on board NAND and requires the OS and file system to be stored on a microSD card. Suggested distributor price for the -xM is $149.

As an Open Source project with a large, enthusiastic community of users, the BeagleBoard would seem like a natural for this book. The basic BeagleBoard is just $5 more than the Mini35. But that's the board only; no cables, no power supply, no display, and no DVD.

Gumstix

The Gumstix series of SBCs are produced by a company of the same name. The name comes from the size of the boards, which is about the size of a stick of chewing gum. There are currently two product lines: the Overo series based on the TI OMAP processors and the Verdex Pro series based on the Marvell Xscale processors.

Owing to its small size, the computer board itself does not have any standard I/O connectors. Connection to the board is through two small 70-pin AVX connectors that mate with an expansion board that provides the usual connections.

Specifications

- Processor:
    - Overo—OMAP3503 or OMAP3530 at 600 or 720 MHz
    - Verdex—PXA270 at 400 or 600 MHz
- 256 or 512 MB of RAM on Overo, up to 128 MB on Verdex
- 256 MB of NAND flash on Overo, up to 32 MB on Verdex
- Bluetooth and 802.11G wireless available as options
- Onboard MicroSD slot

Several expansion boards are available with a wide range of connection options including Ethernet, USB, stereo audio in and out, camera interface, and DVI video, among others. The computer modules range in price from $115 to $229 and the expansion boards go for $27.50–$129.

Raspberry Pi

In March of 2012, the big buzz in the world of ARM-based SBCs is the Raspberry Pi, a $35 board that runs standard Linux distributions such as Fedora and Debian. Developed in the United Kingdom by the Raspberry Pi Foundation, the objective is to provide a low-cost computer that will stimulate the teaching of computer science in schools.

The device went on sale February 29, 2012 and almost immediately the initial production run of 10,000 units sold out. This may ultimately prove to be a suitable alternative to the Mini2440, but at the moment it is apparently very difficult to get.

Specifications

- Processor: Broadcomm BCM2835 at 700 MHz (ARM1176JZFS)
- 256 MB of RAM
- No flash, boots from SD card
- Peripheral connections:

- Ethernet
- USB 2.0
- HDMI
- RCA video
- Stereo audio out
- Header footprint for camera connection
- Powered through microUSB socket
- Size: 85.6 × 53.98 × 17 mm

Setting Up the Mini2440

Figure 5.2 shows the layout of the board. Unfortunately, all of the interesting stuff—the pushbuttons, LEDs, and A/D pot—are hidden by the display panel. Remove the four screws holding the display to the standoffs. There are a couple of alternatives here. You can remove the standoffs and mount them on the other side of the board. Or, you can get four #3 metric nuts to attach the display PCB to the LCD panel and just leave it detached from the main board.

The board is supplied with four cables:

- 9-pin D to 9-pin D RS-232 cable. Note that the cable is female-to-female. You would expect this to be a "null modem" cable where the Tx and Rx wires cross. Not so with this one. It's straight through. So be sure you use this cable.

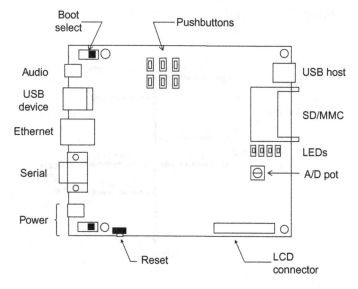

Figure 5.2
Mini2440 Layout.

- Ethernet crossover cable. This allows you to plug the board directly into an Ethernet port on your workstation without having to go through a hub or switch. Think of this as an Ethernet null modem. If you prefer to go through a hub or switch, use a standard straight through cable.
- USB cable. This is needed for initially loading the boot loader. We'll do that in Chapter 6.
- JTAG "wiggler," the 25-pin D to 10-pin flat cable. This is used to drive the JTAG port on the board for the purpose of programming the NOR flash. We'll look at that process in a later chapter.

Flash Memory and File Systems

Flash Memory—NAND and NOR

The Mini2440 has several forms of memory, both volatile and non-volatile, as shown in Figure 5.3. There are two forms of non-volatile storage: a 2 MB NOR flash that occupies part of the processor's memory address space and a 128 MB NAND flash that uses a byte-wide port for address and data. Then there is 64 MB of SDRAM for working memory.

There are two fundamental forms of Flash memory, known as NAND and NOR, named for the basic logic structure of the memory cells. These two types have distinctive characteristics that suit them to different functions in the system.

NOR flash is well suited for code storage because of its reliability, fast read operations, and random access capability. Because codes can be directly executed in place, NOR is ideal

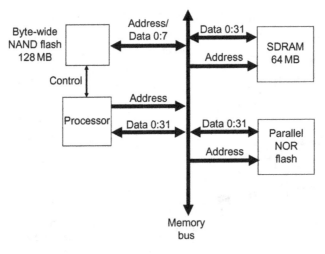

Figure 5.3
Mini2440 Memory Configuration.

for storing firmware, boot code, OSs, and other data that change infrequently. The Mini2440 has 2 MB of NOR flash, although we're not really using it.

The downside to NOR flash is that write and erase times are relatively slow. NAND flash is characterized by higher density, lower cost, faster write and erase times, and a longer rewrite life expectancy. This makes it useful for storing data. But whereas NOR flash is directly accessible in the processor's memory space just like RAM, NAND flash is accessed either by bit serially or a byte at a time.

Our system has 128 MB of byte-wide NAND flash organized as four "partitions":

Name	Size	Offset
u-boot	0×00040000	0×00000000
env	0×00020000	0×00040000
kernel	0×00500000	0×00060000
root fs	$0 \times 07aa0000$	0×00560000

The first partition is 256 KB and holds the boot loader, u-boot. The next partition of 128 KB holds the environment variables for the boot loader. The third partition is the Linux kernel. It occupies 5 MB. Finally, the last partition is everything else and holds the root file system.

Root File System in Flash

In a production-embedded environment, the root file system typically exists as a compressed image in flash memory. When the kernel boots, it copies and uncompresses that image into RAM so that the file system can be writable. One consequence of this is that the file system is *volatile*. Any changes made while the system is powered up are lost when power is removed. To make file changes permanent, you have to reflash the image.

With the root file system stored in NAND flash, we can make changes to files that are reflected directly back to the flash and so are *non-volatile*.

There are many different formats for file systems in Linux, the most popular of which are ext3 and ext4. These file systems generally reside on a mass storage device like a disk and are not well-suited to flash storage devices.

File systems on flash devices must deal with issues such as "wear leveling" and bad blocks. In order to write to a sector of flash memory, it must first be erased. Each sector of a flash device can tolerate a limited number of erase cycles, usually counted in the hundreds of thousands. A file system for flash memory attempts to maximize the useful life of the device by distributing writes evenly over the entire device. Likewise, it detects sectors that have developed flaws and moves data around them.

Ideally, a file system for a flash device will implement these operations transparently so the user or application programmer simply sees the file system as if it were an ext3 or an ext4 residing on a hard disk. Two popular file systems for flash are JFFS2, the Journaling Flash File System version 2, and YAFFS, Yet Another Flash File System. YAFFS is somewhat newer and claims to be better suited to NAND flash devices. The target board does have a YAFFS root file system loaded in NAND flash, but we're not using at the moment because it's not very convenient for software development.

As "journaling" file systems, both JFFS2 and YAFFS can tolerate power loss without corrupting the file system. At worst, the data being written when power fails may be lost.

Preparing the Board

As delivered from the factory, the Mini2440 is not correctly configured for what we want to do. Specifically, it lacks the appropriate boot loader environment for mounting the root file system over Network File System (NFS). There are a number of steps to this process, so read carefully.

Sample Code

On the book's web site (see Section Resources) is a gzipped tar file with several useful items. Download that file and untar it wherever you like, your home directory or perhaps a subdirectory of home. This creates a new directory named EmbeddedLinuxBook/ with several subdirectories under it:

- mkimage—Program to create a u-boot loadable image. Move this to /usr/local/arm/ 4.3.2/bin.
- factory_images/—Contains a number of binary images and scripts for reconfiguring the target.
- home/—Sample source code. This directory will be moved into the target file system under root_qtopia/.
- record_sort/—Sample code for the Eclipse exercise in Chapter 6.

There is already a home/ directory under root_qtopia/. It doesn't have anything useful in it so it's safe to delete. As root user, cd to root_qtopia/ and execute:

```
rm −rf home
```

The r option means remove directories and their contents recursively. The f option means "force," don't ask for confirmation. Now cd to your EmbeddedLinuxBook/ directory and execute:

```
mv home ../root_qtopia
```

The destination shown here assumes EmbeddedLinuxBook/ is directly off your home directory. Finally, execute:

```
mv mkimage/usr/local/arm/4.3.2/bin
```

Make sure the execute bit is set for all users after moving this file. Use the chmod command to set it if necessary.

factory_images

The factory_images/ directory contains:

- README—Describes the files in the directory.
- boot_usb—Program to download a file via USB to the "Supervivi" boot loader in the target's NOR flash.
- mini_boot—U-boot script for setting up the target's NAND flash.
- *.sh—Several shell scripts that simplify some operations.
- s3c2410_boot_usb.tar.gz—Source code for boot_usb.
- set-mini_boot—minicom script for downloading and running mini_boot.
- target_fs.yaffs—YAFFS file system image for loading into NAND flash.
- u_boot.bin—pre-built u-boot image.
- uImage-3.1.5—Executable kernel image, version 3.1.5.
- yaffs.patch—We'll need this file when we configure and build the kernel.

The Script Files

It's worth taking a closer look at the various script files found in factory_images/. One of them you'll need to edit. For each section here, open the corresponding file with an editor and follow along.

mini_boot

This script is interpreted by the u-boot boot loader. It sets up several environment variables with the setenv command and saves the environment to NAND flash. Then it uses Trivial File Transfer Protocol (TFTP) to download three files—the u-boot boot loader, the Linux kernel, and a root file system image—and save them to the appropriate partitions in NAND flash.

The number in line 1, default value 5, identifies the LCD screen connected to the board. The Mini35 version of the Mini2440 ships with any of several different 3.5 in. LCD panels all with different timing and offset parameters. There are also other size display panels. The table below lists all of the displays and the corresponding index number. The panel type is usually identified on the bezel or, if not there, on the bottom of the LCD's PCB.

Index	Designation	Size (inches)	Vendor	Model
0	N35	3.5	NEC	NL2432HC22
1	A70	7.0	Innolux	AT070TN83
2	VGA			VGA shield
3	T35	3.5	Toppoly	TD035STED4
4		5.6	Innolux	AT056TN52
5	X35	3.5	Sony	ACX502BMU
6	W35	3.5	Sharp	LQ035Q1DG06
7	N43	4.3	NEC	NL4827HC19-01B
	N43i		Sharp	LQ043T3DX02

Identify which display your board has and put the corresponding index number in line 1 of `mini_boot`.

Line 2 specifies where the target board's root file system will be found. Replace `<your_home_directory>` with the name of your home directory. Save the file. We won't go into any more detail on this file now because we'll be looking at u-boot in a later chapter.

set-mini_boot

This is a script executed by `minicom`. Its primary role is to download the `mini_boot` script via TFTP and start executing. But before it can do that, it has to set a couple of environment variables for the target's IP address and that of the TFTP server. If you need to use different network addresses, this is the place to change them. Edit lines 2 (target address) and 6 (server address) as appropriate.

*.sh

There are three fairly short Linux shell scripts that encapsulate some details.

`mk_kernel.sh`—The `mkimage` program that adds the u-boot header to a file requires a number of parameters that aren't easy to remember. This script adds the appropriate header to a Linux kernel image.

`mk_uboot.sh`—This script adds the u-boot header to the u-boot script file given by the first argument and moves the resulting file to the TFTP directory, which defaults to `/var/lib/tftpboot` in the absence of a second argument. You need to invoke this script on the file `mini_boot` that you just edited:

```
mk_uboot mini_boot
```

`move_files`—Copies files that will be downloaded to the target to the TFTP directory, `/var/lib/tftpboot` unless an argument is specified. Execute this script.

The Procedure

1. Connect all three cables (USB, network, and serial) and the power supply to the board. Move the boot selector switch to the NOR position, toward the edge of the board.
2. In a Linux shell window, cd to `factory_images/` and execute `minicom -w` (-w for "wrap long lines"). In another shell window, cd to `factory_images/` and become root user.
3. Power up the board. Supervivi presents the menu shown in Figure 5.4. This appears in the window running `minicom`. Press "d". The board is now waiting to download a file.
4. In the other shell window (as root user), execute `./boot_usb u-boot.bin`. The file downloads to the target board and begins execution when the download is finished. U-Boot is now running on the board.
5. In the minicom window, execute the following two commands:

   `nand scrub`   *really* erase the flash

   `nand createbbt`   create a bad block table

   Both of these commands require confirmation that you really want to do this. The `createbbt` command takes a long time. A 1 GB flash can take up to 15 min.
6. In the `minicom` window, type Ctrl-a "g" to bring up the Run a script dialog (Figure 5.5). Type "c" and enter "set-mini_boot". Hit the <Enter> key twice. The `minicom` script runs. This takes a while.

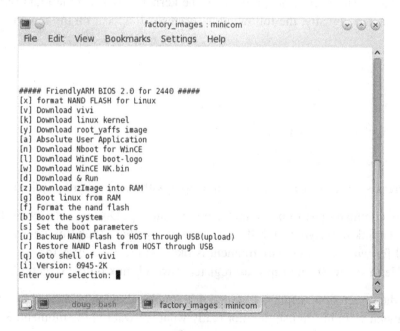

Figure 5.4
Supervivi Menu.

```
+-------------------------[Run a script]-------------------------+
|                                                                |
| A -    Username         :                                      |
| B -    Password         :                                      |
| C -    Name of script   :                                      |
|                                                                |
|     Change which setting?    (Return to run, ESC to stop) █    |
|                                                                |
+----------------------------------------------------------------+
```

Figure 5.5
minicom Run a script Dialog.

7. During the download of the flash file system image, the `minicom` script may "time out," appearing unresponsive. Typing Ctrl-c brings it back to life.
8. When the write of the flash file system image to NAND is complete, move the boot selector switch back to the NAND position (toward the standoff) and press the white reset button. The u-boot boot loader will boot the Linux kernel you just loaded and mount the root file system over NFS from your workstation.

Final Steps

As the kernel is booting up, you'll see a number of messages displayed in the `minicom` (console) window. These track the progress of the kernel's boot. On the LCD screen, you'll see an image of Tux. Finally, the following messages will appear on the screen:

```
Starting networking...
Starting web service...
Starting leds service...
Starting Qtopia, please waiting...
```

The final line in the console window is:

```
Please press enter to activate this console.
```

When you press <Enter> you're logged in as root with no password required.

Execute an `ls` command just to prove that Linux is running. Do `ls -l bin`. Note that almost every entry is a link to busybox. We'll have more to say about BusyBox later in the book. To get a feel for what the target environment is like, list the `/proc` directory and dump the `interrupts` file to see what interrupts are registered with Linux on the target.

Meanwhile, the screen now displays Figure 5.6, a graphical menu generated by Qtopia. Even if the menu were in English, it's not really what we want at the moment. So for now you should disable Qtopia.

Figure 5.6
Qtopia Main Menu.

Qtopia is started from the script file `rcS` in `root_qtopia/etc/init.d`. As root, open the file with an editor and scroll down to the end. Comment out the last three lines by prefacing them with "#". Also comment out lines 44—52 that begin with `/etc/rc.d/init.d/httpd start`. We'll be doing our own thing with a web server and the LEDs and don't want the built-in services interfering.

You may be asking yourself, "How can I open a file as root from a graphical file manager?" There are ways, but many's the time I've hastily opened a file to edit only to realize it's owned by root. The trick I came up with to get around that is to save the file in my home directory using Save as, then become root user in a shell window and move the file back to its original location.

After saving the file, reset the target. It should boot into Linux without the Qtopia screen.

What Can Go Wrong?

It's not unusual to encounter difficulties when bringing up an embedded target board such as the Mini2440. The most common problems fall into two broad categories. The first is the serial port. Make sure the baud rate is set correctly. This is generally not a problem because 115 kBd seems to be the default for minicom. A more common problem is not turning off hardware flow control. Also, be sure you're using the serial cable that came with the kit.

A cable with female connectors on both ends is normally a crossover or "null modem," but the Mini2440 is configured for a straight through cable.

Common networking problems include having Security-Enhanced Linux (SELinux)enabled and/or the firewall turned on. This will prevent the target from NFS mounting its root file system. As a starting point, disable SELinux and turn off the firewall. Later on you can configure either of these features to allow NFS mounting but still provide some level of protection.

Make sure the IP address is correct. And double check that you're using the right network cable—a crossover cable if you're connecting directly between the target and workstation, or a straight through cable if you're going through a hub or switch.

The Boot Loader

The Mini2440 employs a two-stage boot process that we'll explore in detail in Chapter 14. The last stage is an Open Source boot loader called Das U-Boot or just u-boot that resides in NAND flash. Press the reset button, but this time quickly type a key or two. I usually hit the space bar.

Instead of executing its autoboot sequence, u-boot presents its own prompt after printing some basic information about the board.

Enter the command printenv to see the environment variables. There are a lot of them. Here are some of the more significant environment variables:

```
bootdelay = 3
baudrate = 115200
bootcmd = nand read 32000000 kernel 267000; bootm 32000000
stdin = serial
stdout = serial
stderr = serial
ethaddr = 08:08:11:18:12:27
bootargs = console = ttySAC0,115200 root = /dev/nfs nfsroot = 192.168.1.2:/home/doug/
root_qtopia ip = 192.168.1.50
```

Note that the last two lines above are actually one line in U-Boot. The default variables are:

bootdelay—Seconds to wait before proceeding with autoboot.

baudrate—For the serial port.

bootcmd—Command executed if the autoboot sequence is not interrupted.

stdin, stdout, stderr—Direct standard Linux file descriptors to the serial port.

`ethaddr`—Sets the board's MAC address. Note that all Mini2440 boards are delivered with the same default MAC address.

`bootargs`—Command line arguments passed to the Linux kernel when it boots. Information in bootargs includes:

* console device and, because it's a serial port, baud rate,
* root file system type, in this case NFS,
* where the NFS root file system is located,
* the IP address of this device.

Normally, a MAC address is "burned" into an Ethernet device when it is manufactured using numbers obtained from the Institute of Electrical and Electronic Engineers (IEEE) that guarantee that every MAC address in the world is unique. This board vendor chose not to do that, and every board is initially programmed with the same MAC address.

In practice, all that's required is that every MAC address on a particular *network segment* be unique. If you have two or more Mini2440 boards on the same network segment, you'll have to change the MAC address on all but one of them. You can change the address by executing the u-boot command:

```
setenv ethaddr nn.nn.nn.nn.nn.nn
```

where "nn" represents any arbitrary hex digits. If you do change the MAC address or make any other changes to the environment variables, you'll need to execute `saveenv` to make the changes persistent.

U-boot has a very extensive command set, much of which is detailed in Appendix A. For now, type help to see a complete list.

With the target board fully configured and connected to the workstation, we now need a suitable Integrated Development Environment (IDE) to ease the task of writing applications. That's the subject of Chapter 6.

Resources

andahammer.com—The web site for Industrial ARMworks, the US distributor of the Mini2440.
Companion website for this book: http://booksite.elsevier.com/9780124159969.
friendlyarm.net—Chinese manufacturer of the Mini2440. There is an extensive download page, but much of the useful stuff is already on the DVD. The sales page has a complete list of distributors worldwide.
http://billforums.station51.net.
http://groups.google.com/group/mini2440—These are two very good discussion forums on the Mini2440.
intellimetrix.us/downloads.htm—This page also hosts the sample code and any subsequent updates and errata.

Sites for Alternate Boards

BeagleBoard: beagleboard.org.
Gumstix: gumstix.org and gumstix.com.
Raspberry Pi: raspberrypi.org.

## CHAPTER 6

Eclipse Integrated Development Environment[1]

You could spend <u>all day</u> customizing the title bar.
Believe me. I speak from experience.

Matt Welsh

IDEs are a great tool for improving productivity. Desktop developers have been using them for years. Perhaps the most common example is Microsoft's Visual Studio environment. In the past, a number of embedded tool vendors have built their own proprietary IDEs.

In the Open Source world, the IDE of choice is Eclipse, also known as the Eclipse Platform and sometimes just the Platform. The project was started by Object Technology International (OTI), which was subsequently purchased by IBM. In late 2001, IBM and several partners formed an association to promote and develop Eclipse as an Open Source project. It is now maintained by the Eclipse Foundation, a non-profit consortium of software industry vendors. Several leading embedded Linux vendors such as Monta Vista, TimeSys, LinuxWorks, and Wind River Systems have embraced Eclipse as the basis for their future IDE development.

I would go so far as to say that Eclipse is the most professional, well-run Open Source project out there. Every year for about the past 6 years, in the middle of June, the foundation publishes a major new release of the platform and most of the related projects, amounting to something like 20–30 million lines of code. Up until 2011, these releases were named after the moons of Jupiter in alphabetical order. The 2010 release was called Helios. The 2011 release broke with that tradition and is called Indigo.

It should be noted that Eclipse is not confined to running under Linux. It runs just as well under various Windows OSs.

---

[1] Portions of this chapter are adapted from *Eclipse Platform Technical Overview*, © IBM Corporation and The Eclipse Foundation, 2001, 2003, and 2005, and made available under the Eclipse Public License (EPL). The full document is available at www.eclipse.org.

Overview

"Eclipse is a kind of universal tool platform—an open, extensible IDE for anything and nothing in particular. It provides a feature-rich development environment that allows the developer to efficiently create tools that integrate seamlessly into the Eclipse platform," according to the platform's own online overview. Technically, Eclipse itself is not an IDE, but is rather an *open platform* for developing IDEs and *rich client* applications.

Figure 6.1 shows the basic Eclipse workbench. It consists of several *views* including:

- Navigator—shows the files in the user's workspace,
- Text Editor—shows the contents of a file,
- Tasks—a list of "to dos,"
- Outline—of the file being edited. The contents of the outline view are content specific.

Although Eclipse has a lot of built-in functionality, most of that functionality is very generic. It takes additional tools to extend the platform to work with new content types,

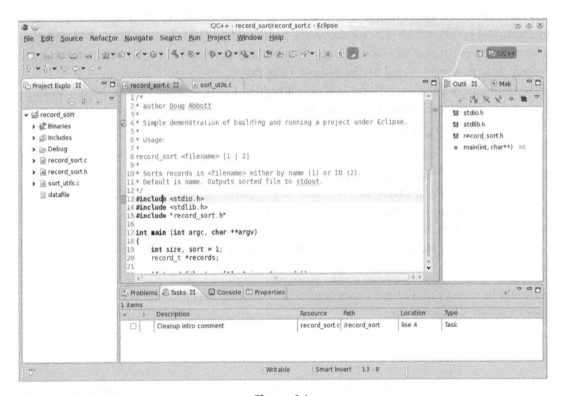

Figure 6.1
The Eclipse Workbench.

to do new things with existing content types, and to focus the generic functionality on something specific.

The Eclipse Platform is built on a mechanism for discovering, integrating, and running modules called *plug-ins*. A tool provider writes a tool as a separate plug-in that operates on files in the workspace and exposes its tool-specific UI in the workbench. When you launch Eclipse, it presents an IDE composed of the set of available plug-ins.

The Eclipse platform is written primarily in Java and in fact was originally designed as a Java development tool. Figure 6.2 shows the platform's major components and APIs. The platform's principal role is to provide tool developers with mechanisms to use, and rules to follow, for creating seamlessly integrated tools. These mechanisms are exposed via well-defined APIs, classes, and methods. The platform also provides useful building blocks and frameworks that facilitate the developing of new tools.

Eclipse is designed and built to meet the following requirements:

- Support the construction of a variety of tools for application development.
- Support an unrestricted set of tool providers, including independent software vendors.
- Support tools to manipulate arbitrary content types such as HTML, Java, C, JSP, EJB, XML, and GIF.
- Facilitate seamless integration of tools within and across different content types and tool providers.
- Support both GUI and non-GUI-based application development environments.
- Run on a wide range of OSs, including Windows and Linux.
- Capitalize on the popularity of the Java programming language for writing tools.

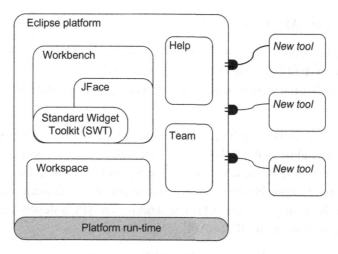

Figure 6.2
Eclipse Platform.

Plug-ins

A *plug-in* is the smallest unit of Eclipse functionality that can be developed and delivered separately. A small tool is usually written as a single plug-in, whereas a complex tool may have its functionality split across several plug-ins. Except for a small kernel known as the Platform Run-time, all of the Eclipse platform's functionality is located in plug-ins.

Plug-ins are coded in Java. A typical plug-in consists of Java code in a JAR library, some read-only files, and other resources such as images, web templates, message catalogs, and native code libraries. Some plug-ins don't contain code at all. An example is a plug-in that contributes online help in the form of HTML pages. A single plug-in's code libraries and read-only content are located together in a directory in the file system, or at a base URL on a server.

Each plug-in's configuration is described by a pair of files. The *manifest* file, manifest.mf, declares essential information about the plug-in, including the name, version, and dependencies to other plug-ins. The second optional file, plugin.xml, declares the plug-in's interconnections to other plug-ins. The interconnection model is simple: a plug-in declares any number of named *extension points* and any number of *extensions* to one or more extension points in other plug-ins.

The extension points can be extended by other plug-ins. For example, the workbench plug-in declares an extension point for user preferences. Any plug-in can contribute its own user preferences by defining extensions to this extension point.

Workbench

The Eclipse *workbench* API, UI, and implementation are built from two toolkits:

1. Standard Widget Toolkit (SWT)—a widget set and graphics library integrated with the native window system but with an OS-independent API.
2. JFace—a UI toolkit implemented using SWT that simplifies common UI programming tasks.

Unlike SWT and JFace, which are both general purpose UI toolkits, the workbench provides the UI personality of the Eclipse Platform, and supplies the structures in which tools interact with the user. Because of this central and defining role, the workbench is synonymous with the Eclipse Platform UI as a whole and with the main window you see when the Platform is running (Figure 6.1). The workbench API is dependent on the SWT API and to a lesser extent on the JFace API.

The Eclipse Platform UI paradigm is based on editors, views, and perspectives. From the user's standpoint, a workbench window consists visually of views and editors. Perspectives manifest themselves in the selection and arrangements of editors and views visible on the screen.

Editors allow you to open, edit, and save objects. They follow an open–save–close lifecycle much like file system–based tools, but are more tightly integrated into the workbench. When active, an editor can contribute actions to the workbench menus and toolbar. The platform provides a standard editor for text resources; other plug-ins supply more specific editors.

Views provide information about some object that you are working with. A view may assist an editor by providing information about the document being edited. For example, the standard content outline view shows a structured outline for the content of the active editor if one is available. A view may augment other views by providing information about the currently selected object. For example, the standard properties view presents the properties of the object selected in another view. The platform provides several standard views; additional ones are supplied by other plug-ins.

A workbench window can have several separate *perspectives*, only one of which is visible at any given moment. Each perspective has its own views and editors that are arranged (tiled, stacked, or detached) for presentation on the screen, although some may be hidden. Several different types of views and editors can be open at the same time within a perspective. A perspective controls initial view visibility, layout, and action visibility. You can quickly switch perspectives to work on a different task, and can easily rearrange and customize a perspective to better suit a particular task. The platform provides standard perspectives for general resource navigation, online help, and team-support tasks. Other plug-ins provide additional perspectives.

Plug-in tools may augment existing editors, views, and perspectives to:

- add new actions to an existing view's local menu and toolbar.
- add new actions to the workbench menu and toolbar when an existing editor becomes active.
- add new actions to the pop-up content menu of an existing view or editor.
- add new views, action sets, and shortcuts to an existing perspective.

Installation

There are two parts to Eclipse: Eclipse itself and the Java Run-time Engine (JRE). Download Eclipse from `eclipse.org`. There are several versions to choose from. Since we're interested in C development, get the Eclipse IDE for C/C++ Developers.

You can put Eclipse wherever you like. I chose to install it in `/usr/local`. After untarring, you'll find a new subdirectory called, not surprisingly, `eclipse/`.

Since Eclipse is written primarily in Java, you'll need a JRE to support it. Most contemporary Linux distributions install a JRE by default, but it may not be compatible with Eclipse. The version supplied with Fedora 14, IcedTea6, works fine. If necessary, download the latest version from Oracle (formerly Sun Microsystems). The download is a binary executable. Copy it wherever you like and execute it. `/usr/local` is a good place. You'll be asked to read and accept the Binary Code License Agreement for the Java SE run-time environment version 6. You'll need to add the `bin/` directory of the newly installed JRE to your PATH.

That's it. Eclipse is installed and ready to go. In a shell window, `cd` to the `eclipse/` directory and execute `./eclipse`. From anywhere else enter the complete path to the eclipse executable. Or just double-click it in a KDE file manager window.

Or better yet, create a script file named `eclipse` in directory `/usr/bin`. It contains the following:

```
#!/bin/bash
/usr/local/eclipse/eclipse &
```

The first line identifies this file as a shell script. The second line executes Eclipse. The "&" says run this program in the background and return the shell prompt. Since `/usr/bin` is in your path, all you have to type is `eclipse`.

Using Eclipse

Every time Eclipse starts up, it asks you to select a *workspace*, a place where your project files are held. The default is a directory called `workspace/` under your home directory. Under `workspace/` then are subdirectories for each project. Following the workspace dialog, if this is the first time you've executed Eclipse, you'll see the Welcome screen. On subsequent runs Eclipse opens directly in the workbench, but you can always access this window from the first item on the Help menu, Welcome.

The Tutorial icon on the Welcome screen leads you to a very good, very thorough introduction to using Eclipse. If you're new to the Eclipse environment, I strongly suggest you go through it. There's no point in duplicating it here. Go ahead, I'll wait.

OK, now you have a pretty good understanding of generic Eclipse operation. It's time to get a little more specific and apply what we've learned to C software development for embedded systems. Since you're running the C/C++ Development Tool (CDT) version of Eclipse, you'll see the C/C++ perspective initially.

The C Development Environment—CDT

Among the ongoing projects at Eclipse.org is the tools project. A subproject of that is the CDT project that has built a fully functional C and C++ IDE for the Eclipse platform. CDT under Linux is based on GNU tools and includes:

- C/C++ Editor with syntax coloring,
- C/C++ Debugger using GDB,
- C/C++ Launcher for external applications,
- Parser,
- Search engine,
- Content assist provider,
- Makefile generator.

Creating a New Project

For our initial exercise in creating a C/C++ project, we'll create a fairly straightforward record sorting application. The records to be sorted consist of a name and an ID number. To simplify the code a bit, we'll replace any spaces in the name field with underscores. The program will sort a file of records in ascending order either by name or ID number as specified on the command line:

```
record_sort <datafile> [1|2]
```

where 1 means sort by name and 2 means sort by ID. Sort by name is the default if no sorting argument is given.

In the Eclipse C/C++ perspective, select `File>New>C Project`. This brings up the New Project wizard as shown in Figure 6.3. Call it "record_sort." The project type is `Executable>Empty Project`; the tool chain is `Linux GCC`, and we'll use the default workspace location. Clicking `Next` brings up the Select Configurations dialog where you can select either or both of the Debug and Release configurations. Later, you'll have the choice of building either of these configurations. The primary difference between them is that the Debug configuration is built with the compiler's debug flag, "-g", turned on to provide symbol information to GDB. The Release configuration leaves out the debug flag.

When you click `Finish` in the New Project wizard, you'll find an entry in the Project Explorer view. The only item under the `record_sort` project is `Includes`, which is a list of paths to search for header files.

At this point, it would be useful to take a look at the directory `workspace/record_sort/`. It contains just two files, `.cproject` and `.project`, both of which are XML codes describing the project. `.project` provides configuration information to the base Eclipse platform while

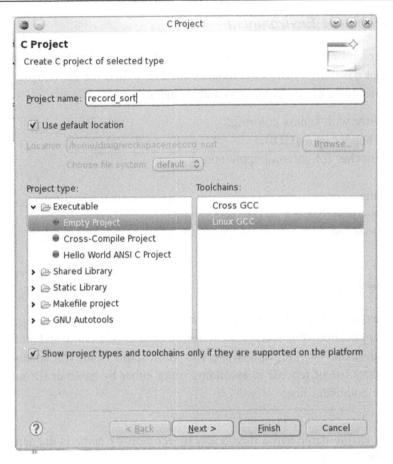

Figure 6.3
New Project Wizard.

the more extensive .cproject file provides information to CDT. It's not necessary to understand the contents of these files, but it is useful to know they exist.

Adding Source Code to the Project

There are basically two ways to add source code to a C project. You can of course create a new file in an Editor window, or you can *import* existing files into the project. Execute File>Import... to bring up the Import Select dialog. Expand the General category and select File System. Click Next, then click Browse, and navigate to EmbeddedLinuxBook/record_sort, and click OK.

This brings up the Import dialog. Select all three files and click Finish. Those files now show up in the Project Explorer view. Note that there is no record_sort.c file. That's because you're going to type it in yourself to get some practice with the CDT editor.

Click the down arrow next to the New icon at the left end of the toolbar and select `Source File` from the drop down menu. Name it "record_sort.c." An Editor window opens up with a preliminary header comment. The contents of `record_sort.c` are given in Figure 6.4, but don't start typing until you read the next section.

Content Assist

The CDT Editor has a number of features to make your coding life easier. These fall under the general heading of *Content Assist*. The basic idea of Content Assist is to reduce the number of keystrokes you must type by predicting what you're likely to type based on the current context, scope, and prefix. Content Assist is invoked by typing `Ctrl + Space` and it's also auto-activated when you type ".", "->", or "::" following a `struct` or `class` name.

Code Templates

Code Templates are an element of Content Assist that provide starting points for frequently used sections of code. When you enter a letter combination followed by `Ctrl + Space`, a list of code templates that begin with that letter combination is displayed. Select one and it is automatically inserted into the file at the current cursor location.

Try this—after entering the `#include` lines in Figure 6.4, type "ma" `Ctrl + Space`. This brings up a template for the `main()` function. You can edit templates at `Window>Preferences>C/C++>Editor>Templates`. Other aspects of Content Assist can also be customized under Preferences.

Automatic Closing

As you type, note that whenever you type an opening quote ("), parenthesis ((), square ([), angle bracket (<), or brace ({), the Editor automatically adds the corresponding closing character and positions the cursor between the two. Type whatever is required in the enclosure and hit `Enter`. This positions the cursor just beyond the closing character. However, if you move the cursor out of the enclosed space, to copy and paste some text for example, the `Enter` key reverts to its normal behavior of starting a new line.

In the case of an opening brace, the closing brace is positioned according to the currently selected coding style and the cursor is properly indented.

Finally, notice that as you type, appropriate entries appear in the Outline view identifying header files, functions, and, if we had any, global variables.

```
/*
 * author Doug Abbott
 *
 * Simple demonstration of building and running a project under
 * Eclipse.
 *
 * Usage:
 *     record_sort <filename> [1 | 2]
 *
 * Sorts records in <filename> either by name (1) or ID (2).
 * Default is name.  Outputs sorted file to stdout.
 */
#include <stdio.h>
#include <stdlib.h>

#include "record_sort.h"

int main (int argc, char **argv)
{
        int size, sort = 1;
        record_t *records;

        if (read_file (argv[1], &size, &records))
        {
                printf ("Couldn't open file %s\n", argv[1]);
                exit (1);
        }
        if (argc > 2)
                sort = atoi (argv[2]);

        switch (sort)
        {
                case 1: sort_name (size, records);
                        break;

                case 2: sort_ID (size, records);
                        break;

                default:
                        printf ("Invalid sort argument\n");
                        return_records (size, records);
                        exit (2);
        }
        write_sorted (size, records);
        return_records (size, records);
        return 0;
}
```

Figure 6.4
record_sort.c.

The Program

Before moving on to building and running the project, let's take a closer look at what it actually does. main() itself is pretty straightforward. It's just a set of calls to functions declared in sort_utils.c that do the real work.

The function `read_file()` reads in a datafile assumed to be organized as one record per line where a record is a text name and a numeric ID. It allocates memory for an array of records and a separate allocation for each name field.

There are two sort functions—one to sort on the name field and the other to sort on the ID field. Both of these implement the shell sort algorithm, named after its inventor Donald Shell. Shell sort improves performance over the simpler insertion sort by comparing elements separated by a gap of several positions.

After the record array is sorted, `write_sorted()` writes it to `stdout`. This of course could be redirected to a file.

The final step is to return all of the allocated memory in the function `return_records()`.

The program does virtually no "sanity checking" and, if you're so inclined, you might want to build some in. There's also very little in the way of error checking.

Building the Project

Once you've completed and saved the `record_sort.c` file, the next step is to build the project. All files that are created in, or imported into, a project automatically become a part of it and are built and linked into the final executable.

In the Project Explorer view, select the top-level `record_sort` entry. Then execute `Project>Build Project` or right-click and select `Build Configurations>Build>All`. In the former case, the *Active Configuration* will be built. By default this is the Debug configuration. The Active Configuration can be changed by executing `Project>Build Configurations>Set Active`.

In the latter case, both the Debug and Release configurations will be built. In either case, one or two new entries will show up under `record_sort` in the Project Explorer view. These entries represent subdirectories called `Debug/` and `Release/` that hold, respectively, the object and executable files for the corresponding build configurations. Each also contains a makefile and some Eclipse-specific files.

Initially, the build will fail because some compile-time errors and warnings have been built into `sort_utils.c`. Open the Problems view, expand the Errors entry, and right-click on the item that says "record_t has no member named 'Id' ." Select Go To to open sort_utils.c, if it isn't already open, and highlight the line that has the error. The correction should be fairly obvious.

Eclipse CDT identifies build problems several different ways. In the Project Explorer view, any project and source files that have problems are flagged with a red X icon for errors or a yellow shield icon with a "!" to indicate a warning. When a source file with errors or warnings is opened in the Editor, the tab shows the same symbol. The Outline view then uses the same icons to show which functions in the file have either errors or warnings.

The Editor window uses the same icons to identify the line on which each error or warning occurs. You can scroll through a source file and click on a line displaying either a warning or error icon and the Problems view will jump to the corresponding entry. If you roll the cursor over a line that's identified as an error or warning, the corresponding error message pops up.

Correct the problems and build the project again. Note incidentally that by default, Eclipse does *not* automatically save modified files before a build. You must manually save them. That behavior can be changed in Preferences. Save sort_utils.c and run the build again. This time it should succeed and you'll see an executable file show up in the Debug tree in the Project Explorer view.

Before moving on the debugging, you might want to turn on line number display in the editor since we'll be referring to line numbers as we go along. In the editor window, right-click in the vertical bar on the left side, called the marker bar. One of the options is Show Line Numbers. Click that. This is a global parameter. It applies to any files currently open in the editor and any that are subsequently opened.

Debugging with CDT

CDT integrates very nicely with the GDB. Like most Unix tools, GDB itself is command line oriented. Eclipse provides a graphical wrapper around the native interface so you don't have to remember all those commands.

Before we can debug, we have to create a *debug launch configuration*. Execute Run>Debug Configurations... Select C/C++ Application and click the New button in the upper left corner. This brings up the Debug Configuration dialog as shown in Figure 6.5. Everything is already filled out correctly. We just need to make one addition. Select the Arguments tab and enter "datafile" into the Program arguments: window. datafile is a set of sample data records for the program to sort. It was imported into the project along with sort_utils.c and record_sort.h.

Click Apply, then Debug. You're asked if you want to open the Debug perspective. Yes, you do.

Figure 6.6 shows the initial Debug perspective. Down at the bottom is a Console window where terminal I/O takes place. The Editor window shows record_sort.c with the first line of main() highlighted. The Outline tab on the right lists all of the header files included by record_sort.c and all of the global and external symbols declared in the file including variables, constants, and functions.

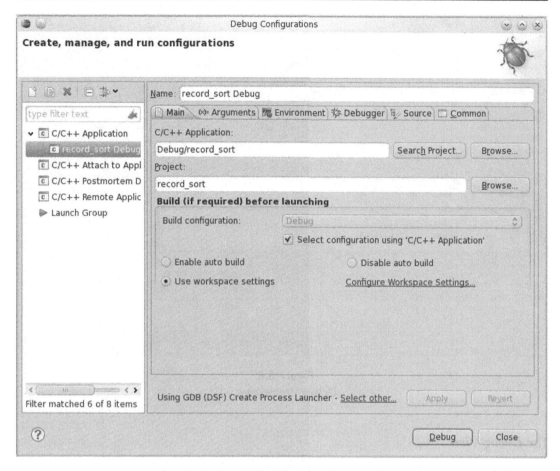

Figure 6.5
Debug Configuration Dialog.

In the upper right is a set of tabs that provide a range of debug information including local variables, breakpoints, and registers. Initially, of course all the local variables have random values since they're allocated on the stack. Just below the Debug tab in the upper left is a set of execution icons. Click on the Step Over icon. The highlight in the Editor window advances to the next line, and sort in the Variable view shows its initial value of 1. The value is highlighted in yellow because it changed since the last time the debugger was suspended. You can also see the current value of any variable by simply rolling the cursor over it. Give it a try.

For now, go ahead and let the program run by clicking the Resume icon in the Debug view toolbar. Hmmm, we didn't get exactly the results we expected. datafile has twelve records, but only one record is output to the Console view. That's because a couple of run-time errors have been built into the program to offer some practice using the debugger.

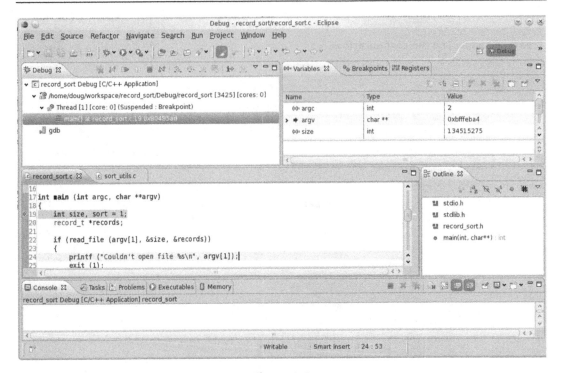

Figure 6.6
Debug Perspective.

The Debug View

In the Debug view, right-click on the top-level project entry and select Relaunch to start
another debug run. The Debug view, shown in Figure 6.7, displays the state of the program
in a hierarchical form. At the top of the hierarchy is a *launch instance*, that is an instance of
a launch configuration identified by its name. Below that is the debugger instance,
identified by the name of the debugger, in this case gdb. Beneath the debugger are the
program threads under the debugger's control. For record_sort there is just one thread.
Later on we'll see how gdb/Eclipse handles multithreaded programs.

Finally, at the lowest level are the thread's stack frames identified by function name, source
code line, and program counter. Currently, there is only one stack frame for main() stopped
at record_sort.c, line 22.

The Debug view's toolbar has lots of buttons for managing program execution. Roll your
cursor over them to see the labels.

Click Step Over once and then click Step Into to get into the read_file() function. Note that
a second stack frame appears in the Debug view and sort_utils.c is opened in the Editor.

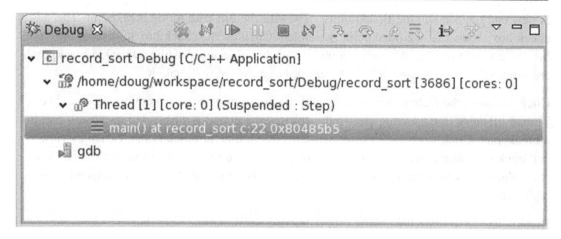

Figure 6.7
Debug View.

At this point it would be worth taking a closer look at the four tabbed views in the upper right of the workbench.

Variables View

When a stack frame is selected in the Debug view, the Variables view displays all the local variables in that frame. Right now you should be in the read_file() function. The two variables visible are both pointers. Clicking the white arrow to the left of the name dereferences the pointer and displays the corresponding value. For string variables, the full string is displayed in the lower window.

Select the main() stack frame in the Debug view and note that the Variables view changes to show the local variables in main(). If anything other than a stack frame is selected, the Variables view goes blank. Remember, you can also view the value of a variable simply by rolling the cursor over it.

Breakpoints View

To debug the problems in record_sort, you'll probably want to set one or more breakpoints in the program and watch what happens. Select the Breakpoints view, which is currently empty because we haven't set any breakpoints.

Let's set a breakpoint at line 34 in sort_utils.c. That's the beginning of an if statement in read_file(). Right-click in the marker bar at line 34 and select Toggle Breakpoint. A green circle appears to indicate that an enabled breakpoint is set at this location. The check mark indicates that the breakpoint was successfully installed.

A new entry appears in the Breakpoints view with the full path to the breakpoint. The check box shows that the breakpoint is enabled. Click on the check box to disable the breakpoint and note that the circle in the marker bar changes to white. Disabled breakpoints are ignored by the debugger. Click the check box again to re-enable it.

Click `Resume` in the Debug view toolbar and the program proceeds to the breakpoint. The Thread [0] entry in the Debug view indicates that the thread is suspended because it had hit a breakpoint. Click `Step Over` to load `temp` with the first record. Select the Variables view and click the white arrow next to `temp`. Now you can see the current values of the fields in `temp`. Variables whose value has changed since the last time the thread was suspended are highlighted in yellow.

Breakpoints have some interesting properties that we'll explore later on.

Memory View

There's one more debug-oriented view that shows up by default in the bottom tabbed window of the Debug perspective. The Memory view lets you monitor and modify process memory. Memory is organized as a list of *memory monitors* where each monitor represents a section of memory specified by a base address. Each memory monitor can be displayed in one or more of four predefined formats known as *memory renderings*. The predefined renderings are hexadecimal (default), ASCII, signed integer, and unsigned integer.

Figure 6.8 shows a memory monitor of the area allocated for `temp` just after the first `fscanf ()` call in `read_file()`. The Memory view is split into two panes, one that lists the currently active monitors and another that displays the renderings for the selected monitor. A monitor may have multiple renderings enabled and these will be tabbed in the renderings pane.

The first four bytes hold a pointer to another area allocated for the name. Remember that the x86 is a "little endian" machine. Consequently, when memory is displayed as shown here, most entries appear to be "backwards."

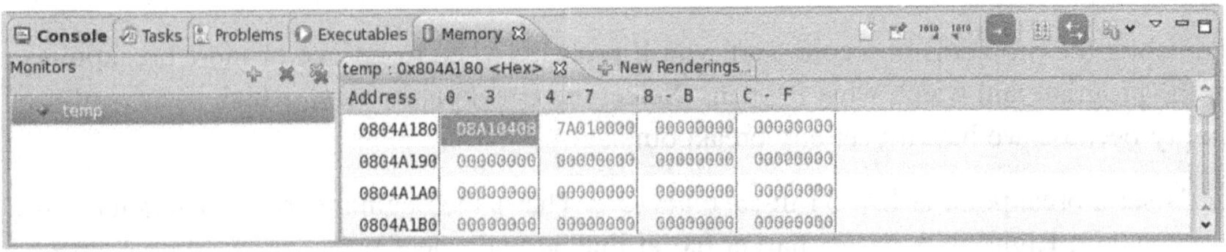

Figure 6.8
Memory View.

Each of the panes has a small, fairly intuitive set of buttons for managing the pane. These allow you to either add or remove monitors or renderings and, in the case of monitors, remove all of them.

Finish Debugging

With this background on the principal debugging features of Eclipse, you should be able to find the two run-time errors that have been inserted in sort_utils.c. Good luck.

Summary

In my humble opinion, Eclipse is the most professionally executed Open Source project I've seen. It's well thought out and meticulously implemented. While it has tremendous power and flexibility, the most common operations are relatively simple and intuitive. The documentation, while still reference in nature and not tutorial, is nevertheless readable, and for the most part accurate. Sadly, there are all too many Open Source projects for which that can't be said.

The plug-in mechanism makes Eclipse infinitely extensible. This opens up opportunities for both Open Source and commercial extensions to address a wide range of applications. The Resources section below lists a repository of Eclipse plug-ins.

Like many of the topics in this book, we've really just skimmed the surface, focusing on those areas that are of specific interest to embedded developers. But hopefully this has been enough to pique your interest in digging further. There's a lot more out there. Go for it!

We're now fully prepared to do some real programming for the target board. That's the subject of the next chapter.

Resources

Abbott, Doug, *Embedded Linux Development Using Eclipse*, Elsevier, 2008. Goes into much more detail about using Eclipse for embedded development.

http://marketplace.eclipse.org/—Eclipse Plugins. This page lists over 1,000 plug-ins, both commercial and Open Source.

www.eclipse.org—The official web site of the Eclipse Foundation. There's a lot here and it's worth taking the time to look through it. I particularly recommend downloading the *C/C++ Development Toolkit User Guide* (PDF).

## PART 2

Application Programming in a Cross-Development Environment

## CHAPTER 7

Accessing Hardware from User Space

The only people who have anything to fear from free software are those whose products are worth even less.

David Emery

Review

This is a good time to stop, take a deep breath and see where we've been before we move on to tackle application development.

Chapter 1 gave us an overview of the embedded and real-time space and how Linux fits into that space. Chapter 2 went through the process of installing a Linux workstation if you didn't already have one. For those new to Linux, Chapter 3 provided an overview and introduction. In Chapter 4 you installed several software packages that we'll need starting with this chapter. Chapter 5 introduced the target SBC that we'll be using throughout this section of the book. At this point, the target should be properly booting Linux. Finally, in Chapter 6 you learned about, and became familiar with, the Eclipse-integrated development environment.

Now it's time to put all that together to build application programs in a cross-development environment.

ARM I/O Architecture

Like most contemporary processor architectures, the ARM places peripheral devices in the same address space as memory. This means that peripherals can be accessed with the same instructions as memory. The program need not know or care whether it's accessing memory or a hardware device. This is in contrast to, for example, the x86 architecture where peripheral devices exist in a separate address space accessed by separate I/O instructions.

In the case of the x86, not only is I/O in a separate address space, the I/O instructions can only be executed from Privilege Level 0, also known as "Kernel space." This necessitates the use of Kernel space device drivers to access peripherals. The ARM, however, lets us access peripherals directly from a user application.

The memory map in Figure 7.1 shows how the 32-bit address space of the S3C2440 processor is allocated. The processor's eight chip select pins each access 128 MB of

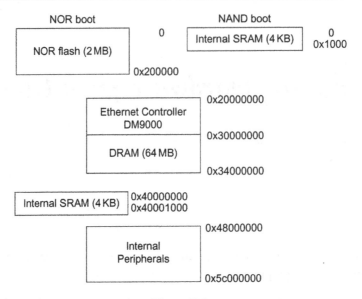

Figure 7.1
S3C2440 Memory Map.

memory space and thus cover the first 1 GB of the 32-bit address space from 0 to 0x3FFFFFFF. This space holds the SDRAM, NOR flash, and Ethernet controller. We'll take up memory addressing later when we discuss the boot process.

Part of the memory mapping is dependent on the setting of the boot select switch. In the NOR position, the NOR flash starts at location 0 and the internal SRAM is mapped to 0x40000000. In the NAND position, the NOR flash is not visible and the internal SRAM is mapped to 0. In this case, the first 4 KB of NAND flash is copied to the internal SRAM and executed. This first stage loader copies U-boot from NAND to SDRAM and executes it.

Oddly, the processor's internal peripherals occupy the address range 0x48000000 to somewhere around 0x5B0000FF. In my experience, peripheral registers usually occupy the top few pages of the address space rather than being "stuck in the middle" like this. The internal peripherals are adequately described in the S3C2440 datasheet on the board's DVD and, with minor exceptions, won't be duplicated here.

LEDs and Pushbuttons

We will however detail the mapping of the four LEDs and six pushbuttons to the general purpose I/O (GPIO) ports. The LEDs use GPIO port B (GPB) based at address 0x56000010 and the pushbuttons use port G (GPG) based at address 0x56000060. Figure 7.2 shows how the LEDs are mapped to bits in the GPB data register and how they appear on the board. Figure 7.3 does the same thing for the pushbuttons.

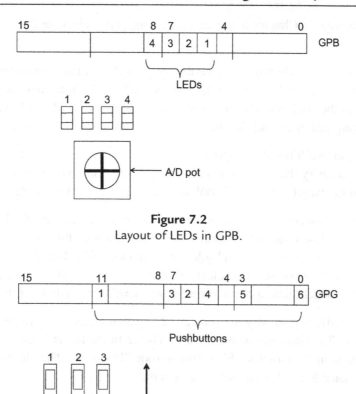

Figure 7.2
Layout of LEDs in GPB.

Figure 7.3
Layout of Pushbuttons in GPG.

As described in the S3C2440 datasheet, each GPIO pin can serve one of several functions as determined by the port's control register. A bit can typically serve as either a general purpose input, general purpose output, or part of a specific peripheral function such as a serial port. Later, we'll see how to program the LED bits as outputs and the pushbutton bits as inputs.

Accessing I/O from Linux—Our First Program

Creating a Project

There are two ways to create a project in Eclipse, a *standard* project, also known as a *Makefile* project or a *managed* project, also known simply as an *Executable*. A managed project automatically creates a project directory, a template C source file, and a makefile in the default workspace. In Chapter 6, we created a managed project for the record_sort program.

Typically, though, you've already got a number of projects put together in the "conventional" way that you'd like to bring into the Eclipse CDT environment, most of the sample projects in the book already have makefiles. The role of a Makefile project then is to bring an existing makefile, and the files they build, into the CDT environment.

For our first project, we'll use the program in root_qtopia/home/src/led[1] to illustrate some basic concepts and verify that CDT is functioning properly and that we can execute a newly built program on the target. Then we'll look at the led program in more detail.

Start up Eclipse and select File -> New -> C Project to bring up the New Project wizard as we did in Chapter 6. Enter the project name, "led," and uncheck the Use default location box. Now browse to home > src > led and select it by clicking OK. In the Project type: window expand Makefile project and select Empty Project. Click Finish. The project wizard creates the project, which then appears in the Project Explorer window on the left.

There's a very specific reason why the project was created under home/ rather than the default workspace. The final executable must be visible to the target board, so we have to locate it somewhere in the target's NFS-mounted root file system. The alternative would be to create a workspace located in the root file system.

The project should automatically build as part of the creation process. If not, right-click the led entry in the Project Explorer view and select Build Project. Okay, we've built the project, now how do we run it?

The Target Execution Environment

Before proceeding, let's review our setup. The target board has a kernel booted from NAND flash and a root file system mounted over the network from the host workstation. stdin, stdout, and stderr on the target are connected to ttySAC0, which in turn is physically connected to ttyS0 or ttyUSB0 on the host. We communicate with the shell running on the target through the terminal emulation program minicom. Figure 7.4 illustrates this setup.

One consequence of mounting the root file system over NFS is that we can execute *on the target* program files that physically reside on the host's file system. This allows us to test the target software without having to program it into flash. And of course, programs running on the target can open files on the NFS-mounted host volume.

One of the directories on the target file system is /home. Under /home/src/ are several subdirectories, each representing a project described later on in this book.

---

[1] From here on, root_qtopia/ will be left off of path names when the usage is unambiguous.

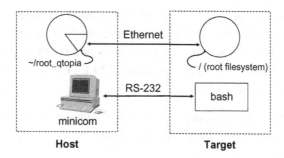

Figure 7.4
The Host and the Target.

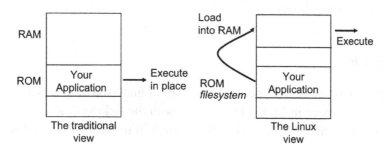

Figure 7.5
Two Views of Embedded Programming.

Start up `minicom` in a shell window and boot the target board. When the boot completes, execute `ls home` in the `minicom` window. You'll see the same files and subdirectories that are in `~/root_qtopia/home` on your workstation. To execute the led program, do:

```
cd home/src/led
./led
```

The program prints a greeting message and then sequentially flashes the four LEDs. Incidentally, the "./" notation in the command represents the current directory. The reason for that is that the current directory is not normally part of a normal user's path. To do so is considered a security risk.

Note that Linux changes the way we have traditionally done embedded programming as shown in Figure 7.5. Embedded systems almost always comprise some combination of RAM and non-volatile memory in the form of ROM, PROM, EPROM, or flash. The traditional way to build an embedded system is to create an executable image of your program, including all library functions statically linked, and perhaps a multitasking kernel. You then load or "burn" this image into one or more non-volatile memory chips. When the system boots, the processor begins executing this image directly out of ROM.

In the Linux view, programs are "files" that must be loaded into memory before executing. Therefore, we create a ROM "file system" containing file images of whatever programs the system needs to run. This may include various utilities and daemons for things like networking. These programs are then loaded into RAM by the boot initialization process, or as needed, and they execute from there.

Generally, the C library is not statically linked to these image files but is dynamically linked so that a single copy of the library can be shared by whatever programs are in memory at a given time.

One of the advantages of the Linux approach is that we're not confined to loading program files from the ROM file system. As just demonstrated, we can just as easily load programs over a network for testing purposes.

The led *Program*

Let's take a closer look at the led program as a way of understanding how to access peripherals from user space in Linux. Open led.c with the Eclipse editor. You'll probably want to turn on line numbers in the editor. Right-click in the marker bar and select Show Line Numbers.

Note first of all the "?" symbols next to all of the #include directives. If you scroll the mouse over one of these, it says "Unresolved inclusion." This means that Eclipse couldn't find the header files and thus can't itself resolve the symbols in the header files. The project will in fact build even though Eclipse reports errors for all the symbols defined by the header files. Also, you can't directly open header files from the Outline view.

This appears to be an artifact of Makefile projects. The managed record_sort project doesn't have this problem. Here's how to fix it:

1. Right-click the project entry in the Project Explorer view and select Properties down at the bottom of the context menu.
2. Expand the C/C++ General entry in the Properties dialog and select Paths and Symbols.
3. Make sure that the Includes tab is selected. Click Add ... and enter /usr/local/arm/ 4.3.2/arm-none-linux-gnueabi/libc/usr/include. Check Add to all configurations and Add to all languages.
4. Click Add ... again and enter ../../include. This picks up the local header file s3c2410-regs.h that defines peripheral registers. Check Add to all configurations and Add to all languages.
5. Click OK. Click Apply. You'll be asked if you'd like to rebuild the index. Yes, you would.

6. Click OK one more time.

The "?" symbols will magically disappear. Note that this is a project-level setting. There does not appear to be a global setting and that probably makes sense. As we'll see later on in this chapter, there's a way to import the settings we just changed into new projects.

Back to the program. On or about line 36 in `main()`, `GPIOp` is declared to be of type `S3C2410P_GPIO`. The latter is a pointer to a structure declared in `s3c2410-regs.h` that maps the GPIO registers in the address space.

Around line 41, the program opens device `/dev/mem`. The `mem` device is a way of referencing memory directly from a user space program. Three lines later we call `mmap()` to map the GPIO register section beginning at 0x56000000 into our process context. Both read and write access to the mapped space are allowed, and it is declared as *shared* meaning that other processes can simultaneously map to the same space.

If `mmap()` succeeds, it returns a pointer, a virtual address, that represents the specified physical space in this process context. The LEDs are attached to bits in parallel I/O controller B (GPB). Before using them, we must configure them correctly.

The LEDs are bits 5–8 of GPB, represented in hex as 0x1E0. The `write_reg()` function at line 49 configures these four bits as outputs. Look at `include/s3c2410-regs.h` to see how this function and the `GP_BIT()` macro work. The `set_reg()` function at line 50 initially sets the bits to 1 meaning the LEDs are off as they are pull downs.

Another useful program is `devmem.c` in the directory `src/devmem/`. It maps a specified region of memory and then lets you read and write locations in that region as bytes, words, or long words. Have a look. The code is fairly self-explanatory.

The Makefile

It's worth taking a look at the project's makefile for a couple of reasons. Open the makefile in the Eclipse editor (Listing 7.1). The first thing to notice is the definition of CC. There are several related conventions for unambiguously naming GNU tool chains. Shown here is the 3-part convention consisting of:

```
<processor>-<operating system>-<tool>
```

Go to `/usr/local/4.3.2/bin` and you'll find a whole set of these 3-part file names that are links to corresponding 5-part file names consisting of:

```
<processor>-<manufacturer>-<operating system>-<executable binary format>-<tool>
```

In this case, which turns out to be common these days, the manufacturer field is set to "none." I've also seen a format that includes the version of the C library. Note however that

```
#
#
#  Makefile for ARM/Linux user application
#

SHELL = /bin/sh

CC = arm-linux-gcc

all: led

led: led.c
        ${CC} -I../../include -g -o $@ $<

clean:
    rm -f *.o
    rm -f led
```

Listing 7.1
Makefile for Led Project.

`arm-linux-g++` and `arm-linux-gcc`, rather than being links, are scripts that invoke the corresponding 5-part file name with an additional parameter that identifies which ARM variant we're compiling for.

A Data Acquisition Example

Now that we have access to the peripheral space of the ARM9 processor, let's move on to something a little more interesting. The S3C2440 incorporates a multichannel A/D converter. Normally, this is used for the touch screen, but for testing purposes, channel 0 is connected to a pot. We'll use that as the basis of a simple data acquisition example that will serve throughout the remainder of this section of the book.

Good programming practice dictates that there be a clear separation between code that is hardware independent and code that actually manipulates I/O registers. At the very least, hardware-dependent code should be in a separate module from the purely application code. In the LED example, we "cheated" somewhat in that the hardware-dependent code was mixed in with the main program in a single source file. In that case it didn't matter much because the objective was to illustrate how hardware access works without introducing any kind of application.

In this example, we'll make a clear distinction between hardware-dependent code and the main program's algorithm. Create a new Eclipse standard C Makefile project named `measure` located at `src/measure`. Open the source file `measure.c`. The first thing to notice is that once again we have those annoying "?" symbols next to the header files. But rather

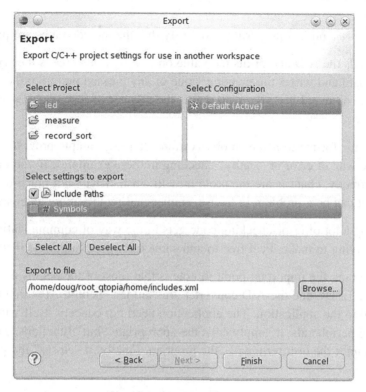

Figure 7.6
Export Dialog.

than laboriously retyping the paths to the headers, this time we'll *export* some project settings from the led project and *import* them into measure.

Proceed as follows:

1.   Select the led project entry in the Project Explorer view. Right-click and select Export. . ..
2.   In the Export dialog, expand C/C++ and select C/C++ Project Settings. Click Next.
3.   Make sure the led project is selected and uncheck # Symbols.
4.   Browse to home/src/ and enter a file name. I used includes. The data will be saved as an XML file.
5.   Figure 7.6 shows the final export dialog. Click Finish.

Next are the steps for importing the project settings into the measure project:

1.   Right-click the measure project entry in Project Explorer and select Import. . ..
2.   Expand C/C++ and select C/C++ Project Settings. Click Next.
3.   Browse to home/src/ and select includes.xml. Click OK.

4. Click Finish.
5. Open the `measure` project properties and verify that the include paths are present.

The program reads the A/D and prints the value on `stdout`. It also reads the six pushbutton inputs on the board and writes the numeric value of any pressed button to the LEDs. The program takes a command line argument that is the number of seconds between readings. The default is two.

This is a good time for me to make an observation about my sample programs. There's very little in the way of error or "sanity" checking in these examples. If you enter something incorrectly, chances are the program will crash. That's not much of a problem in this example but can be a problem in later examples that require more extensive input.

I feel that adding a lot of error-checking code gets in the way of communicating the point the example is trying to make. Feel free to add some error-checking logic if you like.

Back in the program, the important point to note is that most of the details of dealing with the switches, the LEDs, and the A/D converter are "hidden" behind APIs that present data in a form useful to the application. The application need not concern itself with the details of setting up the peripherals; it simply calls the appropriate "init" functions. Then it reads and writes data until the user terminates the program. Finally, it "closes" the peripheral devices.

The file `trgdrive.c` implements the "device driver" APIs. The reason for naming it "trgdrive" will become apparent in Chapter 8. There are two peripheral initialization functions: `initAD()` at line 20 and `initDigIO()` at line 54. Note that `initDigIO()` must be called first because it opens the file descriptor to `/dev/mem` that is also needed by `initAD()`.

`initDigIO()` sets the LED bits of GPB as outputs and the switch bits of GPG as inputs. It returns −1 if it can't open the mem device or can't map the GPIO space.

`initAD()` is equally straightforward. It maps the ADC register space and does a little bit of initialization.

Reading the A/D converter is handled by the generically named `readAD()` function at line 37. It writes the channel number to the ADC and starts the conversion. When conversion is complete, it reads and returns the 10-bit value.

Digital output is embodied in three functions: `setDigOut()`, `clearDigOut()`, and `writeDigOut()`. These functions are specifically adapted to the LEDs on the Mini2440 in that the bit pattern is shifted over so we can use the least-significant bits and "setting" a bit actually clears it to turn on the LED. There's no attempt to verify that the bit mask represents bits configured as output. The input function, `getDigIn()`, simply returns the entire GPG data register. It's up to you to mask the data appropriately.

The two "close" functions simply unmap the memory space that their corresponding "init" functions mapped.

The makefile for the `measure` project has several build targets. By default, Eclipse only knows how to build the "all" target if it is present. In the case of the `measure` project "all" builds something called a "simulation version." We'll talk about that in Chapter 8. Other build targets require that they be declared to Eclipse. On the right-hand side of the C/C++ perspective is a view normally labeled `Outline` that shows an outline of whatever is currently displayed in the Editor window. There is another tab in that view labeled `Make Targets` that allows us to specify alternate make targets. Select that tab, right-click on the `measure` project, and click `Add Target`.

Figure 7.7 shows the dialog brought up by `Add Target`. In this case, the `Target Name` is `measure`. The `Make Target` is also measure, so we don't have to retype it. Eclipse provides a shortcut for the common case when the target name and make target are the same. Click the `create` button. Now in the `Make Targets` view, under the measure project, right-click on the

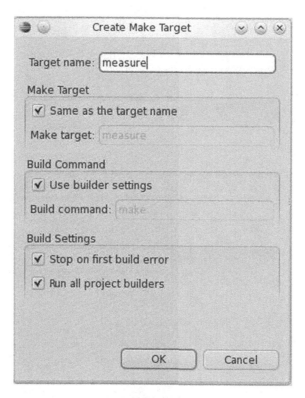

Figure 7.7
Create Make Target Dialog.

measure target and select Build Make Target. In the Project Explorer view on the left-hand side of the perspective, three new files show up: measure.o, trgdrive.o, and measure.

Run the program on the target. The ADC returns 10-bit data in the range of 0–1,023. Turn the pot with a small screwdriver and you should see the reported value change appropriately. Press a pushbutton and notice the pattern in the LEDs.

In this chapter, we learned how to access the peripheral registers on the Mini2440 target board. Chapter 8 will delve more deeply into the subject of debugging.

Resources

DVD: Data Sheets/s3c2440.pdf—Chapter 9 describes the GPIO ports in some detail and Chapter 16 describes the A to D converter.

Debugging Embedded Software

If debugging is the process of removing bugs, then programming must be the process of putting them in.

Edsger W. Dijkstra

In Chapter 6, you saw how Eclipse integrates very nicely with the Gnu Debugger, GDB. In this chapter, we'll look at additional features of GDB and explore how to use GDB on our target board. We'll also consider a simple approach to high-level simulation that can be useful in the early stages of development.

Remote Debugging with Eclipse

In a typical desktop environment, the target program runs on the same machine as the debugger. But in our embedded environment Eclipse with GDB runs on the host workstation and the program being debugged runs on the ARM target. You can think of GDB as having both a client and a server side as illustrated in Figure 8.1. The client is the user interface, and the server is the actual interaction with the program under test. GDB implements a serial protocol that allows the server and the client sides to be separated and to communicate either over an RS-232 link or Ethernet.

There are two approaches to interfacing the target to the gdb serial protocol:

- gdb stubs. A set of functions linked to the target program. gdb stubs is an RS-232-only solution.
- gdbserver. This is a stand-alone program running on the target that, in turn, runs the program to be debugged. The advantage to gdbserver is that it is totally independent of the target program. In other words, the target builds the same regardless of remote debugging. Another advantage of gdbserver is that it runs over Ethernet. Finally, Eclipse uses gdbserver, so that's the best reason of all to use it.

gdbserver is part of the ARM cross tool chain in /usr/local/arm/4.3.2. Interestingly, there are six different copies of gdbserver spread out under arm-none-linux-gnueabi/libc. Through trial and error, I discovered that the one in armv4t/usr/bin works. The gdbserver file must be copied into the target file system so the Mini2440 can execute it. A good location, because it's in the PATH, is /usr/bin.

Linux for Embedded and Real-time Applications.
© 2013 Elsevier Inc. All rights reserved.

DOI: http://dx.doi.org/10.1016/B978-0-12-415996-9.00008-3

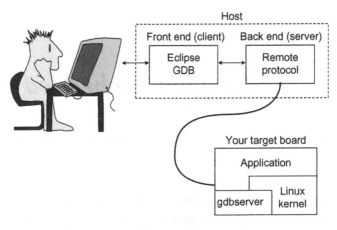

Figure 8.1
Client/Server Nature of GDB.

Then to run the `measure` program under GDB you would first `cd` to the `src/measure` directory and execute:

```
gdbserver:10000 measure
```

The arguments to `gdbserver` are:

- Network port number preceded by a colon. The normal format is "host:port", where "host" is either a host name or an IP address in the form of a dotted quad, but `gdbserver` ignores the host portion. The port number is arbitrary as long as it doesn't conflict with another port number in use. Generally, port numbers below 1024 are reserved for established network protocols such as HTTP (port 80), so it's best to pick a number above 1023. Port number 10000 happens to be the default for the Eclipse debugger.
- Target program name.
- Arguments to the target program, if any.

`gdbserver` responds with something like:

```
Process measure created; pid = 703
```

Your pid value may be different. This says that `gdbserver` has loaded the program and is waiting for a debug session to begin on the specified port.

Remote Debug Launch Configuration

We'll need a different debug launch configuration for remote debugging. With the `measure` project selected in the Project Explorer view, bring up the Debug Configuration dialog either from **Run > Debug Configurations**... or click the down arrow next to the cute little bug icon and select Debug Configurations. ... Even though there's a category called C/C++ Remote Application, we're going to create this new configuration under C/C++ Application just like we did for record_sort.

1. With C/C++ Application selected, click the **New** button. This brings up the Debug Configurations dialog as we saw in Chapter 6 with the name and project entries filled out.
2. We need to fill in the C/C++ Application entry. Click **Search Project**. ... That brings up a dialog listing all of the executables found in the project. Select measure and click **OK**.
3. In the center of the dialog is an entry labeled Build (if required) before launching. I recommend selecting Disable auto build. There will be situations later on where auto build would try to build all the target and fail. The failure is innocuous, but it can be disconcerting.
4. At the bottom of the Main tab is an entry labeled Using GDB (DSF) Create Process Launcher with a link labeled **Select other**. ... Click that link to bring up the dialog shown in Figure 8.2.

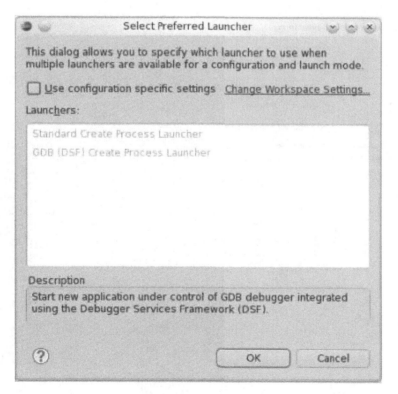

Figure 8.2
Launcher Preference Selection.

5. Check Use configuration specific settings and select Standard Create Process Launcher. Click **OK**.

The issue here is how remote debug sessions are set up and controlled. DSF stands for Debugger Services Framework and offers a more seamless, integrated "experience" that we'll look at later in the chapter. For now, we'll use the simpler approach offered by the Standard Process Launcher.

6. Now select the Debugger tab (Figure 8.3). Since we're debugging ARM code, we need a version of GDB that understands ARM code. And we have to tell it we're debugging remotely.
7. In the Debugger dropdown, select `gdbserver`.
8. Be sure Stop on startup at main is checked.
9. In the GDB Debugger field enter `arm-linux-gdb`.
10. Select the Connection tab under Debugger Options (Figure 8.4). For the Type dropdown, select TCP.

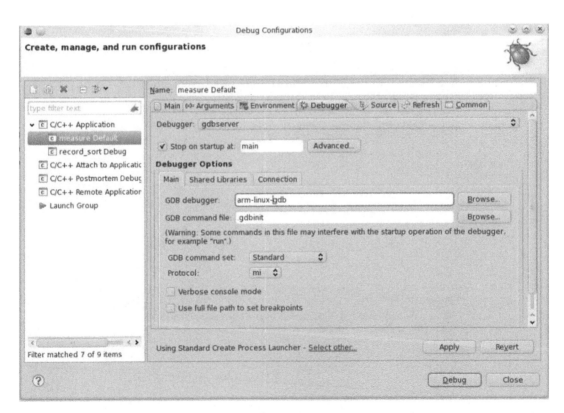

Figure 8.3
Debugger Tab.

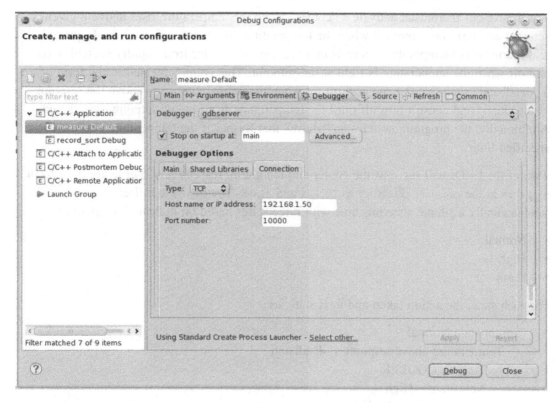

Figure 8.4
Connection Tab.

11. Enter 192.168.1.50 for the Host name or IP address.[1] Now click **Apply**.
12. If `gdbserver` is still waiting patiently on the target board, go ahead and click **Debug**. Otherwise, start up `gdbserver` as described earlier before clicking **Debug**.

Eclipse switches (or asks if you want to switch) to the Debug perspective. Even though you're now debugging on the target board, everything looks pretty much like it did when debugging `record_sort` on the workstation.

A Thermostat

Let's move on to something a little more interesting and potentially very useful. We'll enhance the measure program by turning it into a "thermostat." We'll also take a look at using the host workstation as a preliminary testing environment.

---

[1] "Host" is a little misleading here. It really refers to the target board.

We'll have our thermostat activate a "cooler" when the temperature rises above a given setpoint and turn the cooler off when the temperature falls below the setpoint. In practice, real thermostats incorporate hysteresis that prevents the cooler from rapidly switching on and off when the temperature is right at the setpoint. This is implemented in the form of a "deadband" such that the cooler turns on when the temperature rises above the setpoint + deadband and doesn't turn off until the temperature drops below setpoint—deadband. Additionally, the program includes an "alarm" that flashes if the temperature exceeds a specified limit.

Two of the LEDs will serve as the cooler and the alarm, respectively. These are defined in `driver.h` in the `measure/` directory as `COOLER` and `ALARM`. The thermostat itself is fundamentally a simple state machine with three states based on indicated temperature:

- Normal
- High
- Limit

For each state, the action taken and next state are:

    Current state Normal
        Temperature above setpoint + deadband
            Turn on COOLER
            Next state = High
        Temperature above limit
            Turn on ALARM
            Next state = Limit
    Current state High
        Temperature below setpoint—deadband
            Turn off COOLER
            Next state = Normal
        Temperature above limit
            Turn on ALARM
            Next state = Limit
    Current state Limit
        Temperature below limit
            Turn off ALARM
            Next state = High
        Temperature below setpoint—deadband
            Turn off COOLER
            Next state = Normal

The state machine is just a big `switch()` statement on the state variable.

Here's your chance to do some real programming. Make a copy of measure.c and call it thermostat.c. Since thermostat.c is already a prerequisite in measure's makefile, all you have to do to make it visible to Eclipse is to right-click on measure in the Project Explorer view and select Refresh.

Implement the state machine in thermostat.c. Declare the variables setpoint, limit, and deadband as global integers. I suggest dividing the value returned by readAD() by 10 to get something reasonable as a temperature and to make it easier to adjust the pot. Pick a suitable setpoint and a limit another few "degrees" above that. A deadband of plus/minus one count is probably sufficient.

Host Workstation as Debug Environment

Although remote GDB gives us a pretty good window into the behavior of a program on the target, there are good reasons why it might be useful to do initial debugging on your host development machine. To begin with, the host is available as soon as a project starts, probably well before any real target hardware is available or working. In many cases, it's easier to accurately exercise limit conditions with a simulation than with the real hardware. The host has a file system that can be used to create test scripts and document test results.

When you characterize the content of most embedded system software, you will usually find that something like 5%, maybe 10%, of the code deals directly with the hardware. The rest of the code is independent of the hardware and therefore shouldn't need hardware to test it, provided that the code is properly structured to isolate the hardware-dependent elements.

The idea here is to build a simple simulation that stands in for the hardware I/O devices. You can then exercise the application by providing stimulus through the keyboard and noting the output on the screen. In later stages of testing you may want to substitute a file-based "script driver" for the screen and keyboard to create reproducible test cases.

Take a look at simdrive.c in the measure/ directory. This file exposes the same API as trgdrive.c but uses a shared memory region to communicate with another process that displays digital outputs on the screen and accepts analog input values via the keyboard. This functionality is implemented in devices.c. The shared memory region consists of a data structure of type shmem_t (defined in driver.h) that includes fields for an analog input and a set of digital outputs that are assumed connected to LEDs. It also includes a process ID field (pid_t) set to the pid of the devices process that allows the thermostat process to signal when a digital output has changed.

devices creates and initializes the shared memory region. In simdrive, initAD() attaches to the previously created shared memory region. readAD() simply returns the current value of

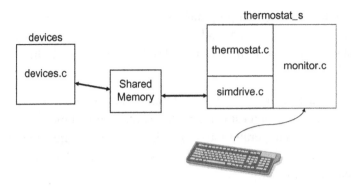

Figure 8.5
Thermostat Simulation.

the a2d field in shared memory. The setDigOut() and clearDigOut() functions modify the leds field appropriately and then signal the devices process to update the screen display. Figure 8.5 illustrates the process graphically.

The executable for the simulation version of the thermostat is called thermostat_s (_s for simulation) and is the default target in the project makefile. In fact, it should have been automatically built when you did the Refresh operation to bring thermostat.c into Eclipse. To build it again after editing thermostat.c, select the Project menu and Build All.

To build the devices program, we need to add another target to the project. Right-click on measure in Project Explorer and select Make targets -> Create. The Target Name is "devices" and the Make Target is the same. Go ahead and build the devices target.

Run devices in a shell window. To debug thermostat_s, we can use the same debug launch configuration we created for the record_sort project in Chapter 6:

1. Bring up the Debug Configurations dialog and select the one for the record_sort project.
2. Change the name to "host." We'll use this configuration for all simulation debugging.
3. Change the project to "measure" and select thermostat_s for the application.
4. Click **Apply**, then click **Debug** to bring up the Debug perspective with the program halted at the first executable line of main().

Advanced Breakpoint Features

The thermostat_s program affords us an opportunity to explore some advanced features of breakpoints. Set a breakpoint at the top of the switch() statement for your state machine. Now go to the Breakpoints view in the upper right of the Debug perspective and right-click

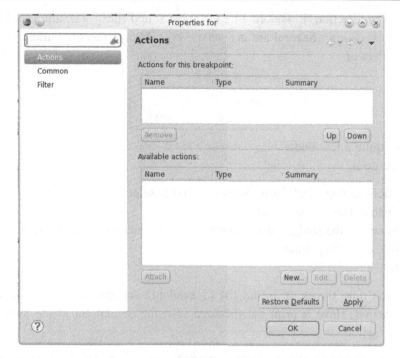

Figure 8.6
Breakpoint Properties.

the breakpoint entry you just created. Select Breakpoint Properties down at the bottom of the context menu. That brings up the rather blank looking dialog of Figure 8.6.

Actions can be attached to a breakpoint such that when it is hit, the attached actions are performed. CDT offers four classes of predefined actions:

- **Play Sound**—Play a selected sound file when the breakpoint is hit. Maybe the breakpoint only happens every half hour or so. You can go off and do something else and when you hear the appropriate beep, you can go back and see what happened. Sound files can be `.wav`, `.mid`, `.au`, or `.aiff`.
- **Log Message**—Output a message to the console. To see the message, you must select Log Action Messages in the Console view.
- **Run External Tool**—Execute a program that has been configured in Eclipse as an external tool. For example, the program might be running on a remote device. You could configure the breakpoint to send an e-mail or SMS to your desktop. Programs are installed and configured from `Run > External Tools > External Tools Configurations...`
- **Resume**—Automatically resume the program after a specified delay. Again, if the program is running remotely, this is probably the only way to keep it running after a breakpoint.

Let's create a sound action. In Fedora 14 KDE most sound files, `.wav`, are found in subfolders of `/usr/share/`. Several are in `sounds/alsa/`. Many others are under `kde4/apps/` in the `sounds/` folders of:

```
kbounce      kmousetool
knetwalk     kolf
korganizer   kreversi
```

There are also quite a few under `/usr/lib/openoffice.org/basis3.3/share/gallery/sounds/`. Here's how we do it:

1. In the Actions dialog, click **New**. Sound Action is selected by default.
2. Give it a name. How about "beep"?
3. Browse to one of the folders that contains `.wav` files and pick one. To hear what it sounds like, click Play Sound.
4. Click **OK**.

The Beep action now shows up in the list of available actions. In a similar fashion, create a Log action. Have it print the message "Hit the breakpoint at the switch statement."

Select each of the entries in the Available actions list and click Attach. They both now appear in the list of Actions for this breakpoint. Let the program run. You should hear the sound when it hits the breakpoint. In the Console view, find the Display Selected Console dropdown near the right-hand side of the Console menu. Select Log Action Messages. To return to program output, select [C/C++ Application] thermostat_s.

Bring up the Breakpoint Properties dialog again. Select Common in the left-hand column to bring up the dialog in Figure 8.7. This lists some basic information about the breakpoint such as the file and line number where it's located and its enabled status. It also offers us a couple of ways of controlling the breakpoint.

Condition lets us specify a Boolean expression that controls whether or not the breakpoint is taken. If the expression evaluates to true the breakpoint is taken. Otherwise it's ignored. The expression can include any program variables. For example, we're only really interested in the breakpoint on the `switch()` statement when we expect the state to change. To take the breakpoint on every pass through the loop would be tedious and unproductive. So we could enter the condition:

```
value > setpoint + deadband
```

Then when the breakpoint is taken, we could step through the code to be sure the state is changed and the cooler is turned on. Then we could change the condition to:

```
value > limit
```

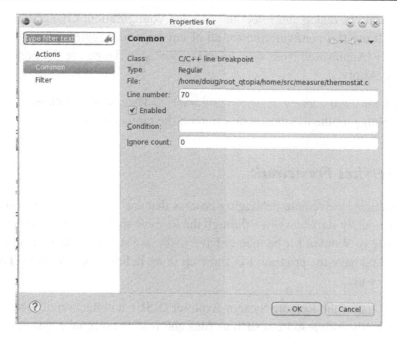

Figure 8.7
Common Breakpoint Properties.

and watch what happens at that transition. Then of course we would reverse the process and watch the state transitions back to normal.

The Ignore count in the Common dialog is the number of times the breakpoint will be ignored before it's taken. For example, if we enter 10, then the first 10 times the breakpoint is encountered, it will be ignored and will finally be taken on the eleventh encounter. This could be useful in a for loop that runs a fixed number of times where we're probably only interested in the first pass through the loop and the last.

Use these advanced breakpoint features to debug your thermostat program by entering different values for the A/D input into the devices program and watch how the thermostat state machine responds.

When you feel the program is running correctly, rebuild it for the target board. You'll need another make target. Call it, for example, ARMthermo. Uncheck Same as the target name and enter all for the Make target. Uncheck Use builder settings and enter make TARGET = 1 for the Build command. Before making this target, delete thermostat.o. Otherwise, it won't be recompiled for the ARM and the linker will complain about a bad object file.

To debug thermostat_t on the target, you'll need to modify the debug launch configuration that you created for the measure program at the beginning of the chapter. Bring up the

Debug Configurations dialog and select the measure configuration. Change the name to "target" as we'll use this configuration for all of our target debugging. Change the file name in C/C++ Application from `measure` to `thermostat_t`. Those are the only changes required. Click **Apply**.

On the target board, run `gdbserver:10000 thermostat_t`. Then back in Eclipse, click **Debug**. You're now debugging the thermostat program on the target board.

Debugger Services Framework

For all its capability, the remote debugging process that we've seen so far is a bit tedious. You have to manually start `gdbserver` through the `minicom` shell and that's where the program console is. Wouldn't it be nice to have `gdbserver` start up automatically when you click **Debug**? And have the program I/O show up in an Eclipse console view. That's what the DSF does for us.

DSF is built on top of the Remote System Explorer (RSE), a collection of tools that allows you to work directly with resources such as files and folders on remote systems. RSE in turn uses SSH, the Secure SHell, to establish communication with remote systems. Unfortunately, our target file system doesn't include SSH. So before we can play around with RSE and DSF, we'll have to install SSH on the target.

Installing SSH

SSH is a network protocol for secure data communication, remote shell services, command execution, and other secure network services between networked computers connected via a secure channel over an insecure network. It is a replacement for Telnet and other insecure remote shell protocols such as the Berkeley rsh and rexec protocols, which send information, notably passwords, in plaintext, rendering them susceptible to interception and disclosure using packet analysis. The encryption used by SSH is intended to provide confidentiality and integrity of data over an unsecured network.

There are a number of implementations of SSH floating around, many of them proprietary. There is also an Open Source version called, perhaps not surprisingly, OpenSSH. This is what we'll use.

Installing OpenSSH on our target board turns out to be relatively easy as someone has done all the work for the Mini2440 and kindly posted the results to a Google group. See the Resources section for the URL. From the top-level group page, go to Source and then Browse. You'll see a list of five files, a README, a script, and three gzipped tar files. SSH is built on top of two other packages—the secure sockets layer (SSL) and zlib, a cryptographic library. Download all five files to some convenient place on your workstation.

Take a look at the README.txt file. It outlines three major steps: run the script to build the packages, copy files to the appropriate locations in the target file system, and run the SSH daemon to verify that it works. The first step says to run sh build.sh. It's not entirely clear to me why you need to explicitly run sh rather than just execute ./build.sh. The man page for sh implies that it is an alternate way of invoking bash and goes on to say "If bash is invoked with the name sh, it tries to mimic the startup behavior of historical versions of sh as closely as possible, while conforming to the POSIX standard as well." As an experiment, I executed ./build.sh and it seemed to build correctly, so the choice is up to you.

Open build.sh with an editor. The first thing to notice is the series of rm commands starting at line 12. You probably don't have any of those directories at this point, so you could safely comment out those lines. build.sh reveals a pattern that is very common in building Open Source packages. The first step is make clean that deletes all of the intermediate files created by a previous build, if there was any. The next step is configure, a script that first checks the system on which it is executing to be sure that it is capable of building the package. Then it creates the Makefile and anything else necessary to build the package based on the system it found and any parameters passed into the script. The configure script often takes arguments that modify the way the package is built.

The make step actually builds the package using the files created by the configure script as the "recipe." The output of make is a set of files, libraries, and/or executables, in the current directory, or subdirectories thereof. These will typically need to be moved to some other location to be useful. That's the role of the make install command. In many cases, make install must be run as root user because the directories it references have root privileges. In this case, depending on where you chose to load the files in the first place, you may not need to be root.

You will, however, need to be root to carry out the rest of the steps in README.txt because the target file system is largely owned by root. Step 2 of README.txt is really a continuation of the make install process but couldn't be folded into the Makefile or a script because the author doesn't know the location of your target file system. You are instructed to copy a couple of files to usr/libexec. Turns out that directory doesn't currently exist in our target file system, so you'll have to create it first. On the other hand, usr/local/etc does exist, so you don't have to create that one.

The var/ directory is interesting. Open root_qtopia/etc/init.d/rcS in an editor. This is the script that is executed when Linux first boots on the target board. For our purposes, the interesting part starts about line 30 with /bin/mount −n −t ramfs none /var. This says mount a RAM file system at /var. /var represents "variable" data that tend to change a lot, things like log files. If our root file system were mounted on the NAND flash, having the /var directory as part of that, would result in unnecessary wear on the flash memory chips as well as decreased performance resulting from writing log files to flash. So the /var directory is mounted on a file system that resides in RAM. Among other things, this means the contents of /var do not persist over a reboot.

Where this is going is that README.txt instructs you to create a couple of directories under /var. /var/run and /var/empty that already exist because they're created by the rcS script. What you need to do is add the line mkdir –p /var/log/empty/sshd to rcS. You will also have to add the line /usr/sbin/sshd to rcS to get the SSH daemon to start up at boot time. Reboot the target and verify that sshd is running.

Add a Password for Root

When SSH connects to a remote system, it insists that you enter a password even if the user you're logging in as doesn't require a password, which is the case with the root user on our target board. You do this with the passwd utility. Execute passwd on the target board. It responds:

```
Changing password for root
New password:
```

Then it will prompt you to verify the password. Now whenever you boot the board you will have to enter the password.

Configuring RSE

Fire up Eclipse and open the RSE perspective. You'll find that the Project Explorer view on the left has been replaced by a view labeled Remote Systems, and the collection of views under the Editor has been replaced by something called Remote System Details. Right now the only thing listed in Remote Systems is Local, your workstation. Expand that entry to reveal Local Files and Local Shells. Expand Local Files to browse the file system (Figure 8.8).

Right-click on Local Shells and select Launch Shell. This brings up a Remote Shell view in the lower tabbed window. Commands are entered in the small window at the bottom and the results show up in the larger window. Figure 8.9 shows the last few lines of ls root_qtopia from the home directory on my system. I find the presentation of this shell a little awkward and hard to use. There's a more useful shell format available in the remote shell once it is set up.

Needless to say, the RSE isn't very interesting until we connect it to a remote system, specifically the Mini2440 target board. In the Remote Systems view, click the **Define a**

Figure 8.8
Remote Systems View.

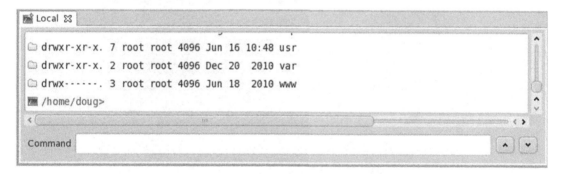

Figure 8.9
Local shell view

connection icon in the upper left-hand corner or right-click Local and select **New - >
Connection**. The first step is to select the Remote System Type. In our case, it will be SSH
Only.

Why not the Linux type, you may ask. That uses a different, RSE-specific communication
protocol called dstore, which has more functionality but requires more resources on the

server. dstore is in fact Java-based, so it requires a Java Runtime Environment (JRE) running on the server and for the moment that just seems like too much trouble.

Before clicking **Next** to bring up the dialog in Figure 8.10, be sure your target board is powered up into Linux. Host name defaults to LOCALHOST. Change it to the IP address of your target board. Note that the Connection name is initially the same as the Host name, but you can change it to anything you want, like "Mini2440" for example. Likewise the Description can be anything.

Clicking **Next** shows the file services available on the remote machine. There's nothing here that needs to be changed. Clicking **Next** again shows the available shell services. Again, nothing needs changing. Clicking **Next** one more time shows the available terminal services. Click **Finish** and the new connection shows up in the Remote Systems view.

Expand the new connection to reveal entries similar to what we saw with Local. When you expand My Home or Root under Sftp Files, you are required to enter a valid user ID (root)

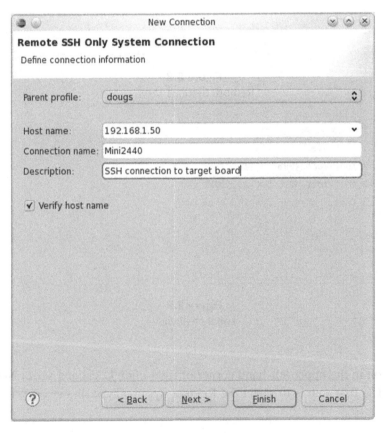

Figure 8.10
Setting Up an SSH Connection.

and password for the remote system, which effectively logs you into it. You can now use copy and paste commands to move files between the local host and the remote system. You can also open remote text files in the Eclipse editor by double-clicking them.

Even this isn't particularly useful in the present circumstances where the target's file system is mounted from the workstation over NFS. We have direct access to the file system from the workstation. But if the target were truly a remote computer, it would definitely be useful.

Note the entry Ssh Terminals under the Mini2440 connection. Right-click that and select Launch Terminal. This brings up another form of remote shell as shown in Figure 8.11. This one looks more like the shell that we've been using through minicom and I personally find it easier to use than the so-called shell view available with the local connection. Incidentally, at this point we can do away with minicom and the serial port for communicating with Linux. The only reason we need a serial port any more is for communicating with the u-boot boot loader. But for our purposes, the real value of RSE is in debugging.

Debugging with RSE

We can now set up a debug configuration to use the SSH connection to the target. Bring up the Debug Configurations dialog, select C/C++ Remote Application, and click the **New launch configuration** icon. That brings up the dialog in Figure 8.12, shown with the final values. The top part of the Main tab looks like what we've seen before, selecting the Application and the Project. Again, it's probably best to select Disable auto build.

In the Connection dropdown, select the remote connection that you just created. Next, we have to tell GDB where to find the executable on the remote target. In most cases, RSE would have to "download" the executable in order to execute and debug it on the remote machine. In our case of course, the executable is already visible to the target in the NFS-mounted file system, so you can check Skip download to target path at the bottom of the

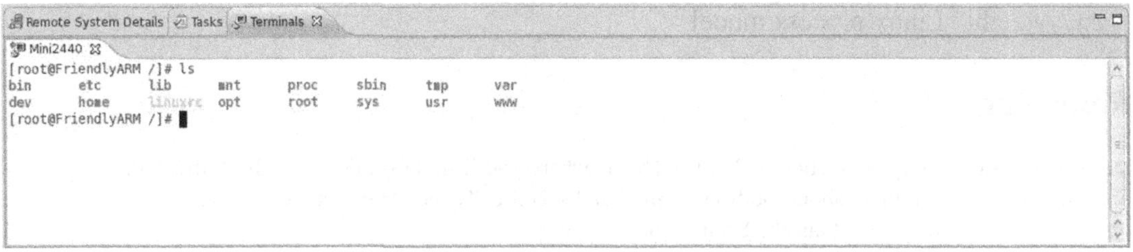

Figure 8.11
Ssh Terminal View.

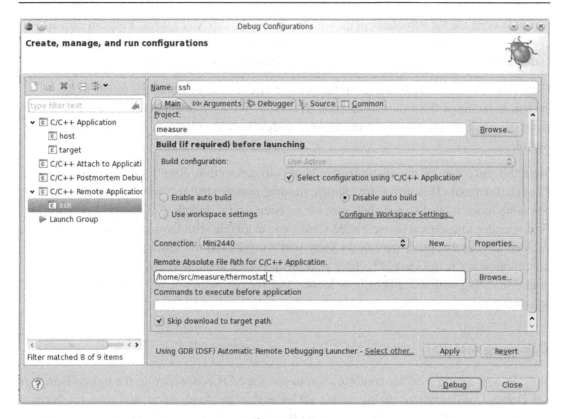

Figure 8.12
Remote Debug Configuration.

Main tab. If you do choose to let RSE download the executable, don't use the current location because it will try to overwrite itself and fail. Try /home/src.

Click **Apply** and then **Debug**. After some delay, the Debug perspective will appear, ready to debug thermostat_t on the target board. Debugging proceeds just as before but now the program output appears in an Eclipse console view.

In the next chapter, we'll look at multithreaded programming as an alternative to the "heavyweight" Linux process model.

Resources

http://code.google.com/p/openssh4mini2440/—This is where you'll find OpenSSH for the Mini2440.
openssh.org—To learn more about the Secure Shell and specifically the Open Source version.
openssl.org—To learn more about the Secure Sockets Layer.

## CHAPTER 9

Posix Threads

Linux is only free if your time has no value.

Jamie Zawinski

The thermostat that we developed in the last chapter isn't very practical because the operating parameters, setpoint, limit, and deadband are hardcoded into the program. Any practical thermostat would make these parameters user adjustable. In this case, we might invent a simple command protocol to change the parameters through the console.

How might we implement such a process? Back in the old DOS days, we might have used the function kbhit() to poll the keyboard for activity. But Linux is a multitasking system. Polling is tacky. What we need is an independent thread of execution that waits for a line of text on the console, parses the line, and acts on it.

We could use fork() to create a new process. Listing 9.1 illustrates in pseudocode form a possible implementation. We start by creating a shared memory space and then fork to create a new child process. The child process monitors stdin and posts the result of any valid command to the appropriate entry in the shared memory region. The parent is the thermostat program as we've already seen it.

But there's probably a more efficient implementation. Remember from Chapter 3 that a Linux process can have multiple threads of execution, and all of these threads share the process's memory space. So how about we create a thread that just sits on stdin waiting for a command? Specifically, in this chapter we'll explore Posix threads.

Posix, also written as POSIX, is an acronym that means Portable Operating System Interface with an X thrown in for good measure. POSIX represents a collection of standards defining various aspects of a portable OS based on UNIX. These standards are maintained jointly by the Institute of Electrical and Electronic Engineers (IEEE) and the International Standards Organization (ISO). The various documents have been pulled together into a single standard in a collaborative effort between the IEEE and The Open Group (see the Resources section).

In general, Linux conforms to Posix. The command shell, utilities, and system interfaces have all been upgraded over the years to meet Posix requirements. But in the context of

Linux for Embedded and Real-time Applications.
© 2013 Elsevier Inc. All rights reserved.

DOI: http://dx.doi.org/10.1016/B978-0-12-415996-9.00009-5

```
#include <unistd.h>
#include "measure.h"

int running = 1;
params_t *p; //pointer to shared memory

int main (int argc, void **argp)
{
        create shared memory space;
        switch(fork())
        {
          case -1:
              printf ("fork failed\n");
              break;

          case 0:  // child process
              attach to shared memory space;
              while (running)
              {
                fgets();
                parse command;
                putresultin shared memory;
              }
              break;

          default: // parent process
              attach to shared memory space;
              while (running)
              {
                read A/D;
                act on current state;
              }
              break;
        }
        exit (0);
}
```

Listing 9.1
Fork Implementation of Thermostat.

multithreading, we are specifically interested here in the Posix Threads interface known as
1003.1 c.

The header file that prototypes the Pthreads API is `pthread.h` and resides in the usual
directory for library header files, `/usr/include`, or in the case of our cross tool chain, `/usr/`
`local/arm/4.3.2/arm-none-linux/gnueabi/libc/usr/include`.

Threads

RTOSs often refer to threads as tasks. They're pretty much the same thing. It is an
independent thread of execution embodied in a function. The thread has its own stack,
referred to as its *context*.

```
int pthread_create (pthread_t *thread, pthread_attr_t *attr, void
     *(* start_ routine)(void *),void *arg);
void pthread_exit (void *retval);
int pthread_join (pthread_t thread, void**thread_return);
pthread_t pthread_self (void);
int sched_yield (void);
```

A thread is created by calling `pthread_create()` with the following arguments:

- `pthread_t`—A *thread object* that represents or identifies the thread. `pthread_create()` initializes this as necessary.
- Pointer to a thread *attribute* object. Often it is NULL. More on this later.
- Pointer to the *start routine*. The start routine takes a single pointer to void argument and returns a pointer to void.
- Argument to be passed to the start routine when it is called.

A thread may terminate by calling `pthread_exit()` or simply returning from its start function. The argument to `pthread_exit()` is the start function's return value.

In much the same way that a parent process can wait for a child to complete by calling `waitpid()`, a thread can wait for another thread to complete by calling `pthread_join()`. The arguments to `pthread_join()` are the thread object of the thread to wait on and a place to store the thread's return value. The calling thread is blocked until the target thread terminates. There is no parent/child relationship among threads as there is with processes, so a thread can join any other thread.

A thread can determine its own ID by calling `pthread_self()`. Finally, a thread can voluntarily yield the processor by calling `sched_yield()`.

Note that most of the functions above return an int value. This reflects the Pthreads approach to error handling. Rather than reporting errors in the global variable `errno`, Pthreads functions report errors through their return value. Appendix B gives a more complete description of the Pthreads API including a list of all error codes.

Thread Attributes

POSIX provides an open-ended mechanism for extending the API through the use of *attribute objects*. For each type of Pthreads object, there is a corresponding attribute object. This attribute object is effectively an extended argument list to the related object create or initialize function. A pointer to an attribute object is always the second argument to a create function. If this argument is NULL, the create function uses appropriate default values. This also has the effect of keeping the create functions relatively simple by leaving out a lot of arguments that normally take default values.

An important philosophical point is that all Pthreads objects are considered to be "opaque." Most of them anyway. We'll see an exception shortly. This means that you never directly access members of the object itself. All access is through API functions that get and set the member arguments of the object. This allows new arguments to be added to a Pthreads object type by simply defining a corresponding pair of get and set functions for the API. In simple implementations, the get and set functions may be nothing more than a pair of macros that access the corresponding member of the attribute data structure.

```
int pthread_attr_init (pthread_attr_t *attr);
int pthread_attr_destroy (pthread_attr_t *attr);
int pthread_attr_getdetachstate (pthread_attr_t *attr, int *detachstate);
int pthread_attr_setdetachstate  (pthread_attr_t *attr, int detachstate);

Scheduling Policy Attributes
int pthread_attr_setschedparam (pthread_attr_t *attr, const struct sched_param *param);
int pthread_attr_getschedparam (const pthread_attr_t *attr, struct sched_param *param);
int pthread_attr_setschedpolicy (pthread_attr_t *attr, int policy);
int pthread_attr_getschedpolicy (constpthread_attr_t *attr, int*policy);
int pthread_attr_setinheritsched (pthread_attr_t *attr, int inherit);
int pthread_attr_getinheritsched (const pthread_attr_t *attr, int *inherit
```

Before an attribute object can be used, it must be initialized. Then any of the attributes defined for that object may be set or retrieved with the appropriate functions. This must be done before the attribute object is used in a call to pthread_create( ). If necessary, an attribute object can also be "destroyed." Note that a single attribute object can be used in the creation of multiple threads.

The only required attribute for thread objects is the "detach state." This determines whether or not a thread can be joined when it terminates. The default detach state is PTHREAD_CREATE_JOINABLE meaning that the thread can be joined on termination. The alternative is PTHREAD_CREATE_DETACHED, which means the thread can't be joined.

Joining is useful if you need the thread's return value or you want to make sure threads terminate in a specific order. Otherwise it's better to create the thread detached. The resources of a joinable thread can't be recovered until another thread joins it whereas a detached thread's resources can be recovered as soon as it terminates.

There are also a number of optional scheduling policy attributes.

Synchronization—Mutexes

As soon as we introduce a second independent thread of execution, we create the potential for resource conflicts. Consider the somewhat contrived example of two threads, each of which wants to print the message "I am Thread n" on a single shared printer as shown in

Figure 9.1. In the absence of any kind of synchronizing mechanism, the result could be something like "II a amm TaTasskk 12."

What is needed is some way to regulate access to the printer so that only one task can use it at a time.

A *mutex* (short for "mutual exclusion") acts like a key to control access to a resource. Only the thread that has the key can use the resource. In order to use the resource (in this case a printer), a thread must first *acquire* the key (mutex) by calling an appropriate kernel service (Figure 9.2). If the key is available, that is, the resource (printer) is not currently in use by someone else, the thread is allowed to proceed. Following its use of the printer, the thread releases the mutex so another thread may use it.

If, however, the printer is in use, the thread is blocked until the thread that currently has the mutex releases it. Any number of threads may try to acquire the mutex while it is in use.

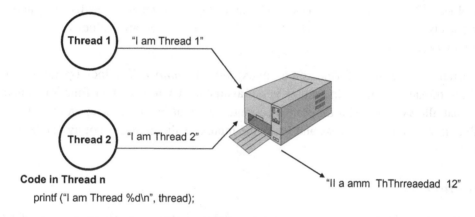

Code in Thread n
printf ("I am Thread %d\n", thread);

Figure 9.1
Resource Conflict.

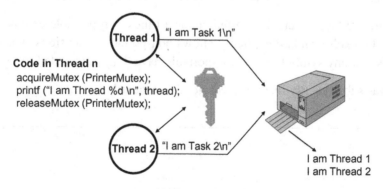

Code in Thread n
acquireMutex (PrinterMutex);
printf ("I am Thread %d \n", thread);
releaseMutex (PrinterMutex);

Figure 9.2
Solving Resource Conflict with Mutex.

All of them will be blocked. The waiting threads are queued either in order of priority or in the order in which they called `pthread_mutex_lock()`. The choice of how threads are queued at the mutex may be built into the kernel or it may be a configuration option when the mutex is created.

```
pthread_mutex_t mutex = PTHREAD_MUTEX_INITIALIZER;

int pthread_mutex_init (pthread_mutex_t *mutex, const pthread_mutexattr_t *mutex_attr);
int pthread_mutex_destroy (pthread_mutex_t *mutex);

int pthread_mutex_lock (pthread_mutex_t *mutex);
int pthread_mutex_unlock (pthread_mutex_t *mutex);
int pthread_mutex_trylock (pthread_mutex_t *mutex);
```

The Pthreads mutex API follows much the same pattern as the thread API. There is a pair of functions to initialize and destroy mutex objects and a set of functions to act on the mutex objects. The listing also shows an alternate way to initialize statically allocated mutex objects. `PTHREAD_MUTEX_INITIALIZER` provides the same default values as `pthread_mutex_init()`.

Two operations may be performed on a mutex: *lock* and *unlock*. The lock operation causes the calling thread to block if the mutex is not available. There's another function called *trylock* that allows you to test the state of a mutex without blocking. If the mutex is available, *trylock* returns success and locks the mutex. If the mutex is not available, it returns `EBUSY`.

Mutex Attributes

Mutex attributes follow the same basic pattern as thread attributes. There is a pair of functions to create and destroy a mutex attribute object. We'll defer discussion of the pshared attribute until later. There are some other attributes we'll take up shortly.

The mutex attribute "type" actually started out as a Linux non-portable extension to Pthreads. The Pthreads standard explicitly allows non-portable extensions. The only requirement is that any symbol that is non-portable have "_np" appended to its name.

Mutex type was subsequently incorporated into the standard.

```
int pthread_mutexattr_init (pthread_mutexattr_t *attr);
int pthread_mutexattr_destroy (pthread_mutexattr_t *attr);

int pthread_mutexattr_gettype (pthread_mutexattr_t *attr, int *type);
int pthread_mutexattr_settype (pthread_mutexattr_t *attr, int type);
```

```
int pthread_mutexattr_gettype (pthread_mutexattr_t *attr, int *type);
int pthread_mutexattr_settype (pthread_mutexattr_t *attr, int type);

type =    PTHREAD_MUTEX_NORMAL
          PTHREAD_MUTEX_ERRORCHECK
          PTHREAD_MUTEX_RECURSIVE
          PTHREAD_MUTEX_DEFAULT

int pthread_mutexattr_getprioceiling (const pthread_mutexattr_t *mutex_attr, int *prioceiling);
int pthread_mutexattr_setprioceiling (pthread_mutexattr_t *mutex_attr, int prioceiling);
int pthread_mutexattr_getprotocol (const pthread_mutexattr_t *mutex_attr, int *protocol);
int pthread_mutexattr_setprotocol (pthread_mutexattr_t *mutex_attr, intprotocol);

protocol =PTHREAD_PRIO_NONE
          PTHREAD_PRIO_INHERIT
          PTHREAD_PRIO_PROTECT
```

What happens if a thread should attempt to lock a mutex that it has already locked? Normally, the thread would simply hang up. The "type" attribute alters the behavior of a mutex when a thread attempts to lock a mutex that it has already locked. The possible values for type are:

Normal. If a thread attempts to lock a mutex it already holds, it is blocked and thus effectively deadlocked. The normal mutex does no consistency or sanity checking.
Error checking. If a thread attempts to recursively lock an error-checking mutex, the lock function returns immediately with the error code EDEADLK. Furthermore, the unlock function returns an error if it is called by a thread other than the current owner of the mutex.
Recursive. A recursive mutex allows a thread to successfully lock a mutex multiple times. It counts the number of times the mutex was locked and requires the same number of calls to the unlock function before the mutex goes to the unlocked state. This type also checks that the mutex is being unlocked by the same thread that locked it.
Default. The standard says that this type results in "undefined behavior" if a recursive lock is attempted. It also says that an implementation may map Default to one of the other types.

Optionally, a Pthreads mutex can implement the priority inheritance or priority ceiling protocols to avoid priority inversion. The mutex attribute *protocol* can be set to "none," "priority inheritance," or "priority ceiling," The *prioceiling* attribute sets the value for the priority ceiling.

Problems with Solving the Resource Sharing Problem—Priority Inversion

Using mutexes to resolve resource conflicts can lead to subtle performance problems. Consider the scenario illustrated in Figure 9.3. Threads 1 and 2 each require access to a

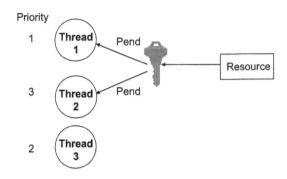

Figure 9.3
Priority Inversion Scenario.

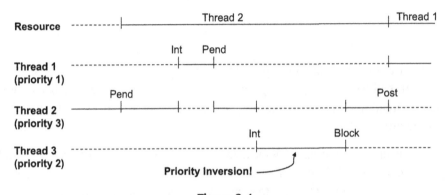

Figure 9.4
Execution Timeline.

common resource protected by a mutex. Thread 1 has the highest priority and Thread 2 has the lowest. Thread 3, which has no need for the resource, has a "middle" priority.

Figure 9.4 is an execution timeline of this system. Assume Thread 2 is currently executing and pends on the semaphore. The resource is free so Thread 2 gets it. Next, an interrupt occurs that makes Thread 1 ready. Since Thread 1 has higher priority, it preempts Thread 2 and executes until it pends on the resource mutex.

Since the resource is held by Thread 2, Thread 1 blocks and Thread 2 regains control. So far everything is working as we would expect. Even though Thread 1 has higher priority, it simply has to wait until Thread 2 is finished with the resource.

The problem arises if Thread 3 should become ready while Thread 2 has the resource locked. Thread 3 preempts Thread 2. This situation is called *priority inversion* because a lower priority thread (Thread 3) is effectively preventing a higher priority thread (Thread 1) from executing.

A common solution to this problem is to temporarily raise the priority of Thread 2 to that of Thread 1 as soon as Thread 1 pends on the mutex. Now Thread 2 can't be preempted by anything of lower priority than Thread 1. This is called *priority inheritance*.

Another approach, called *priority ceiling*, raises the priority of Thread 2 to a specified value higher than that of any task that may pend on the mutex as soon as Thread 2 gets the mutex. This is considered to be more efficient because it eliminates unnecessary context switches. No thread needing the resource can preempt the thread currently holding it.

Posix threads has optional attributes for setting a mutex's protocol as either priority inheritance or priority ceiling and for setting the priority ceiling value.

Communication—Condition Variables

There are many situations where one thread needs to notify another thread about a change in status to a shared resource protected by a mutex. Consider the situation in Figure 9.5 where two threads share access to a queue. Thread 1 reads the queue and Thread 2 writes it. Clearly, each thread requires exclusive access to the queue and so we protect it with a mutex.

Thread 1 will lock the mutex and then see if queue has any data. If it does, Thread 1 reads the data and unlocks the mutex. However, if the queue is empty, Thread 1 needs to block somewhere until Thread 2 writes some data. Thread 1 must unlock the mutex before blocking or else Thread 2 would not be able to write. But there's a gap between the time Thread 1 unlocks the mutex and blocks. During that time, Thread 2 may execute and not recognize that anyone is blocking on the queue.

The condition variable solves this problem by waiting (blocking) with the mutex locked. Internally, the conditional wait function unlocks the mutex allowing Thread 2 to proceed. When the conditional wait returns, the mutex is again locked.

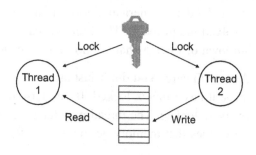

Figure 9.5
Communicating via Condition Variable.

```
pthread_cond_t cond = PTHREAD_COND_INITIALIZER;

int pthread_cond_init (pthread_cond_t *cond, const pthread_condattr_t *cond_attr);
int pthread_cond_destroy (pthread_cond_t *cond);

int pthread_cond_wait (pthread_cond_t *cond, pthread_mutex_t *mutex);
int pthread_cond_timedwait (pthread_cond_t *cond, pthread_mutex_t *mutex, const struct time-
        spec *abstime);
int pthread_cond_signal (pthread_cond_t *cond);
int pthread_cond_broadcast (pthread_cond_t *cond);
```

The basic operations on a condition variable are *signal* and *wait*. Signal wakes up one of
the threads waiting on the condition. The order in which threads wake up is a function of
scheduling policy. A thread may also execute a *timed wait* such that if the specified time
interval expires before the condition is signaled, the wait returns with an error. A thread
may also *broadcast* a condition. This wakes up all threads waiting on the condition.

Condition Variable Attributes

Pthreads does not define any required attributes for condition variables although there is at
least one optional attribute.

Thread Termination and Cancellation

A thread may be terminated either voluntarily or involuntarily. A thread terminates itself either
by simply returning or by calling `pthread_exit()`. In the latter case, all *cleanup handlers* that
the thread registered by calls to `pthread_cleanup_push()` are called prior to termination.

Most threads run in an infinite loop. As long as the system is powered up, the thread is
running, doing its thing. Some threads start up, do their job, and finish. But there are also
circumstances where it's useful to allow one thread to terminate another thread
involuntarily. Perhaps, the user presses a CANCEL button to stop a long search operation.
Maybe the thread is part of a redundant numerical algorithm and is no longer needed
because another thread has solved the problem. The Pthreads cancellation mechanism
provides for the orderly shutdown of threads that are no longer needed.

But cancellation must be done with care. You don't just arbitrarily stop a thread at any
point in its execution. Suppose it has a mutex locked. If you terminate a thread that has
locked a mutex, it can never be unlocked. The thread may have allocated one or more
dynamic memory blocks. How does that memory get returned if the thread is terminated?

Pthreads allows each thread to manage its own termination such that the thread can free up
and/or return any global resources before it actually terminates. So when you cancel a

Mode	State	Type	Meaning
Off	Disabled	N/A	Cancellation remains pending until enabled
Deferred	Enabled	Deferred	Cancellation occurs at next cancellation point
Asynchronous	Enabled	Asynchronous	Cancellation may occur at any time

Figure 9.6
Cancellation Modes.

thread you're usually not stopping it immediately, you're asking it to terminate itself as soon as it's safe or convenient.

Pthreads supports three cancellation modes (Figure 9.6) encoded as two bits called *cancellation state* and *cancellation type*. A thread may choose to disable cancellation because it is performing an operation that must be completed. The default mode is Deferred meaning that cancellation can only occur at specific points, called *cancellation points*, where the program tests whether the thread has been requested to terminate. Most functions that can block for an unbounded time, such as waiting on a condition variable or reading or writing a file, should be cancellation points and are defined as such in the Posix specification.

While asynchronous cancellation mode might seem like a good idea, it is rarely safe to use. That's because, by definition you don't know what state the thread is in when it gets the cancellation request. It may have just called pthread_mutex_lock(). Is the mutex locked? Don't know. So while asynchronous cancellation mode is in effect, you can't safely acquire any shared resources.

```
int pthread_cancel (pthread_t thread);
int pthread_setcancelstate (int state, int *oldstate);
int pthread_setcanceltype (int type, int *oldtype);
void pthread_testcancel (void);
```

A thread can cancel another thread by calling pthread_cancel(). pthread_setcancelstate() and pthread_setcanceltype() allow a thread to set its cancellation mode. Note that these functions return the previous value of state and type, respectively. The function pthread_testcancel() allows you to create your own cancellation points. It returns immediately if the thread has not been requested to terminate. Otherwise it doesn't return.

Cleanup Handlers

When a thread is requested to terminate itself, there may be some things that need to be "cleaned up" before the thread can safely terminate. It may need to unlock a mutex or return a dynamically allocated memory buffer, for example. That's the role of *cleanup handlers*. Every thread conceptually has a stack of active cleanup handlers. Handlers are pushed on the stack by `pthread_cleanup_push()` and executed in reverse order when the thread is canceled or calls `pthread_exit()`. A cleanup handler takes one `void*` argument.

```
void pthread_cleanup_push (void (*routine)(void *), void *arg);
void pthread_cleanup_pop (int execute);
```

The most recently pushed cleanup handler can be popped off the stack with `pthread_cleanup_pop()` when it's no longer needed. Often the functionality of a cleanup handler is needed whether or not the thread terminates. The execute argument specifies whether or not a handler is executed when it's popped. A non-zero value means execute. Note also that `pthread_cleanup_pop()` can only be called from the same function that called `pthread_cleanup_push()`. Aside from being good programming practice, this is necessary because `pthread_cleanup_push()` is a macro that, in many implementations, ends with an opening brace, "{", introducing a block of code. `pthread_cleanup_pop()` then has the corresponding closing brace.

Listing 9.2 shows the read thread (the main function) of a typical queuing application with a cleanup handler added. We assume the default deferred cancellation mode. Note that `pthread_cleanup_pop()` is used to unlock the mutex rather than the normal mutex unlock function.

The reason we need a cleanup handler here is that `pthread_cond_wait()` is a cancellation point, and the mutex is locked when we call it. But is it really necessary to push and pop the cleanup handler on every pass through the while loop? It is if there is a cancellation point in the section called "do something with the data" where the mutex is unlocked. This thread can only invoke the cleanup handler if it has the mutex locked. If there are no cancellation points while the mutex is unlocked then it's safe to move the push cleanup call outside the loop. In that case we don't really need pop cleanup.

We'll have an opportunity to use a cleanup handler in the next chapter.

Pthreads Implementations

Until kernel version 2.6, the most prevalent threads implementation was LinuxThreads. It has been around since about 1996 and by the time development began on the 2.5 kernel, it was generally agreed that a new approach was needed to address the limitations in LinuxThreads. Among these limitations, the kernel represents each thread as a separate process, or *schedulable entity*, giving it a unique process ID, even though many threads exist within one process entity. This causes compatibility problems with other thread implementations. There's a hard-coded limit of 8192 threads per process, and while this

```
/* Cleanup Handler Example */
#include <pthread.h>

typedef struct my_queue_tag {
  pthread_mutex_t  mutex;        /* Protects access to queue */
  pthread_cond_t   cond;         /* Signals change to queue */
  int              get, put;     /* Queue pointers */
  unsigned char    empty, full;  /* Status flags */
  int              q[Q_SIZE]     /* the queue itself */
} my_queue_t;

my_queue_t data = {
  PTHREAD_MUTEX_INITIALIZER, PTHREAD_COND_INITIALIZER, 0, 0, 1, 0};

void cleanup_handler (void *arg)
/*
  Unlocks the mutex associated with a queue
*/
{
  pthread_mutex_unlock (mutex *) arg);
}

int main (int argc, char *argv[])
{

  while (1)
  {
    pthread_cleanup_push (cleanup_handler, (void *) &data.mutex);
    pthread_mutex_lock (&data.mutex);
    if (queue is empty)
    {
    data.empty = 1;
    pthread_cond_wait (&data.cond, &data.mutex);
    }
    /* read data from queue */
    pthread_cleanup_pop (1);

    /* do something with the data */
  }
}
```

Listing 9.2
Example of a cleanup handler.

may seem like a lot, there are some problems that can benefit from running thousands of threads.

The result of this new development effort is the Native Posix Threading Library or NPTL, which is now the standard threading implementation in 2.6 and 3.x series kernels. It too treats each thread as a separately schedulable entity but takes advantage of improvements in the kernels that were specifically intended to overcome the limitations in LinuxThreads. The clone() call was extended to optimize thread creation. There's no longer a limit on the number of threads per process, and the new fixed time scheduler can handle thousands of threads without excessive overhead. A new synchronization mechanism, the Fast Mutex or "futex," handles the non-contention case without a kernel call.

In tests on an IA-32, NPTL is reportedly able to start 100,000 threads in 2 s. By comparison, this test under a kernel without NPTL would have taken around 15 min.

Upgrading the Thermostat

We now have enough background to add a thread to our thermostat to monitor the serial port for changes to the parameters. We'll use a very simple protocol to set parameters consisting of a letter followed by a space followed by a number. "s" represents the setpoint, "l" the limit, and "d" the deadband. So to change the setpoint you would enter, for example,

```
s 65<Enter>
```

This sets the setpoint to 65°.

Copy `thermostat.c` that currently resides in `src/measure/` to `src/posix/`. Create a new Eclipse project called "posix" located in the `home/src/posix/` directory. Again, be sure it's a makefile project and import the project settings with the include paths. Open `monitor.c`. The first thing to notice is `#include <pthread.h>`. This header file prototypes the Pthreads API. Note the declarations of `paramMutex` and `monitorT`. The latter will become the handle for our monitor thread.

Because the parameters, `setpoint`, `limit`, and `deadband`, are accessed independently by the two threads, the thermostat and monitor, we need a mutex to guarantee exclusive access to one thread or the other. In reality, this particular application probably doesn't absolutely require the mutex because the shared resources are integers and access to them should be atomic. But it does serve to illustrate the point.

Note incidentally that `setpoint`, `limit`, and `deadband` are declared `extern` in `thermostat.h`. It is assumed that these variables are allocated in `thermostat.c`. If you happened to use different names for your parameters, you'll need to change them to match.

Take a look at the `monitor()` function. It's just a simple loop that gets and parses a string from `stdin`. Note that before `monitor()` changes any of the parameters, it locks `paramMutex`. At the end of the switch statement it unlocks `paramMutex`. This same approach will be required in your thermostat state machine.

Move down to the function `createThread()`. This will be called from `main()` in `thermostat.c` to initialize `paramMutex` and create the monitor thread. `createThread()` returns a non-zero value if it fails. `terminateThread()` makes sure the monitor thread is properly terminated before terminating `main()`. It should be called from `main()` just before `closeAD()`.

Linux Device Drivers

We're going to introduce another Linux "feature" with this exercise. Instead of accessing the I/O ports directly from User space, we'll make use of Kernel space device drivers that further isolate us from the hardware. Here we focus on using drivers from User space. In Chapter 12 we'll look at the drivers themselves.

While other OSs treat devices as files, "sort of," Linux goes one step further in actually creating a directory for devices. Typically this is /dev. One consequence of this is that the redirection mechanism can be applied to devices just as easily as to files.

Devices come in four "flavors": Character, Block, Pipe, and Network. The principal distinction between character and block is that the latter, such as disks, are randomly accessible, i.e., you can move back and forth within a stream of characters. With character devices, the stream generally moves in one direction only. Block devices are accessed through a file system whereas character devices are accessed directly. In both cases, I/O data are viewed as a "stream" of bytes.

Pipes are pseudodevices that establish unidirectional links between processes. One process writes into one end of the pipe and another process reads from the other end.

Network devices are different in that they handle "packets" of data for multiple protocol clients rather than a "stream" of data for a single client. This necessitates a different interface between the kernel and the device driver. Network devices are not nodes in the /dev directory.

From here on out we'll only be dealing with character devices.

The Low Level I/O API

The set of User Space system functions closest to the device drivers is termed "low level I/O." For the kind of device we're dealing with in this chapter, low level I/O is probably the most useful because the calls are inherently synchronous. That is, the call doesn't return until the data transfer has completed.

```
int open (const char *path, int oflags);
int open (const char *path, int oflags, mode_t mode);
size_t read (int filedes, void *buf, size_t count);
size_t write (int filedes, void *buf, size_t count);
int close (int filedes);
int ioctl (int filedes, int cmd, …);
```

The box shows the basic elements of the low level I/O API.

- OPEN. Establishes a connection between the calling process and the file or device. `path` is the directory entry to be opened. In the case of a peripheral device, it is usually an entry in the `/dev` directory. `oflags` is a bitwise set of flags specifying access mode and must include one of the following:

O_RDONLY   Open for read-only.

O_WRONLY   Open for write-only.

O_RDWR   Open for both reading and writing.

Additionally `oflags` may include one or more of the following modes:

O_APPEND   Place written data at the end of the file.

O_TRUNC   Set the file length to zero, discarding existing contents.

O_CREAT   Create the file if necessary. Requires the function call with three arguments where `mode` is the initial permission.

If OPEN is successful it returns a non-negative integer representing a "file descriptor." This value is then used in all subsequent I/O operations to identify this specific connection.
- READ and WRITE. These functions transfer data between the process and the file or device. `filedes` is the file descriptor returned by OPEN. `buf` is a pointer to the data to be transferred and `count` is the size of `buf` in bytes. If the return value is non-negative, it is the number of bytes actually transferred, which may be less than `count`.
- CLOSE. When a process is finished with a particular device or file, it can close the connection, which invalidates the file descriptor and frees up any resources to be used for another process/file connection. It is good practice to close any unneeded connections because there is typically a limited number of file descriptors available.
- IOCTL. This is the "escape hatch" to deal with specific device idiosyncrasies. For example, a serial port has a baud rate and may have a modem attached to it. The manner in which these features are controlled is specific to the device. So each device driver can establish its own protocol for the IOCTL function.

A characteristic of most Linux system calls is that, in case of an error, the function return value is −1 and doesn't directly indicate the source of the error. The actual error code is placed in the global variable `errno`. So you should always test the function return for a negative value and then inspect `errno` to find out what really happened. Or better yet call `perror()`, which prints a sensible error message on the console.

There are a few other low level I/O functions but they're not particularly relevant to this discussion.

The Linux kernel provides a collection of three device drivers to access the peripheral devices that are relevant to our particular application. These are:

- `/dev/adc`
  - `read()` returns a numeric text string for channel 0.
- `/dev/leds`
  - `ioctl()` sets or clears individual LEDs.
- `/dev/buttons`
  - `read()` returns a 6-character string; "1" or "0" for each button.

Take a look at `trgdrive.c` in the `posix/` directory to see how these drivers are used in the Posix thermostat.

The drivers themselves are beyond the scope of this chapter, but if you'd like to peruse them, they're in `linux/drivers/char` as:

```
mini2440_adc
mini2440_buttons
mini2440_leds
```

We'll be looking at device drivers and kernel modules in Chapter 12.

Changes Required in `thermostat.c`

Surprisingly, little changes to make `thermostat.c` work with the monitor.

1. `include pthread.h` and `thermostat.h`.
2. call `createThread()` and test the return value at the top of `main()`.
3. lock `paramMutex` just before the switch statement on the state variable and unlock it at the end of the switch statement.
4. call `terminateThread()` just after the `running` loop.

That's it! The makefile supports both target and simulation builds as described in Chapter 8. The simulation build uses the driver file `simdrive.o` from the `measure/` directory and assumes that it's built. Build the simulation and try it out.

Debugging Multithreaded Programs

Multithreading does introduce some complications to the debugging process but, fortunately, GDB has facilities for dealing with those. Debug `thermostat_s` in Eclipse (be sure `devices` is running) using the host debug configuration and set a breakpoint at the call to `createThread()` in `main()`. When it stops at the breakpoint, click Step into to go into the `createThread()` function.

Step over the pthread_mutex_init() call and CHECK_ERROR. Now when you step over the call to pthread_create(), a second thread appears in the Debug view. This is the monitor thread and it's already inside the fgets() call.

Set a breakpoint in the monitor() function just after the call to fgets(). Now enter a line of text in the Console view. It doesn't really matter if it's a valid command or not. All we want to do at this point is get the program to stop at the breakpoint. Let the program continue, and when it reaches the breakpoint note that Thread [2] is now the one suspended at a breakpoint. Thread [1] is most likely suspended inside the sleep() function.

Play around with setting breakpoints in various places in both main() and monitor() to verify that parameters get updated properly and the thermostat responds correctly to the new values.

When you're satisfied with the program running under the simulation, rebuild it for the target and debug it there. You'll need to create a new make target just as we did in the previous chapter for the measure project. You'll also need to delete both monitor.o and thermostat.o for the target version to build properly.

When you start the debugger you're likely to see some warning messages whiz by in the Console view. These are innocuous.

The debugger stops at the first line of main() as normal. Let the program execute to the breakpoint at createThread(). Step over and the Debug view shows two threads just as we saw in the simulation version. From here on out, the target debugging process is essentially identical to the simulation mode.

With a basic understanding of threads, we're ready to tackle networking, the subject of the next chapter.

Resources

The Open Group has made available free for private use the entire Posix (now called Single Unix) specification. Go to:

www.unix.org/online.html

You'll be asked to register. Don't worry, it's painless and free. Once you register you can read the specification online or download the entire set of html files for local access.

This chapter has been little more than a brief introduction to Posix threads. There's a lot more to it. An excellent, more thorough treatment is found in:

Butenhof, David R., *Programming with POSIX Threads*, Addison-Wesley, 1997.

## CHAPTER 10

Embedded Networking

Give a man a fish and you feed him for a day.
Teach him to use the Net and he won't bother you for weeks.

Everything is connected to the Internet, even refrigerators.[1] So it's time to turn our attention to network programming in the embedded space. Linux, as a Unix derivative, has extensive support for networking.

We'll begin by looking at the fundamental network interface, the socket. With that as a foundation, we'll go on to examine how common application-level network protocols can be used in embedded devices.

Sockets

The "socket" interface, first introduced in the Berkeley versions of Unix, forms the basis for most network programming in Unix systems. Sockets are a generalization of the Unix file access mechanism that provides an endpoint for communication either across a network or within a single computer. A socket can also be thought of as an extension of the named pipe concept that explicitly supports a client/server model wherein multiple clients may be attached to a single server.

The principal difference between file descriptors and sockets is that a file descriptor is bound to a specific file or device when the application calls open() whereas sockets can be created without binding them to a specific destination. The application can choose to supply a destination address each time it uses the socket, for example, when sending datagrams, or it can bind the destination to the socket to avoid repeatedly specifying the destination, for example, when using TCP.

Both the client and the server may exist on the same machine. This simplifies the process of building client/server applications. You can test both ends on the same machine before distributing the application across a network. By convention, network address 127.0.0.1 is a "local loopback" device. Processes can use this address in exactly the same way they use other network addresses.

---

[1] http://www.engadget.com/2011/01/12/samsung-wifi-enabled-rf4289-fridge-cools-eats-and-tweets-we-go/ or just Google "internet refrigerator". I don't suppose there are any net-connected toaster ovens yet.

Linux for Embedded and Real-time Applications.
© 2013 Elsevier Inc. All rights reserved.

DOI: http://dx.doi.org/10.1016/B978-0-12-415996-9.00010-1

Try it Out

Execute the command /sbin/ifconfig. This will list the properties and current status of network devices in your system. You should see at least two entries: one for eth0, the Ethernet interface, and the other for lo, the local loopback device.

This command should work on both your development host and your target with similar results. ifconfig, with appropriate arguments, is also the command that sets network interface properties.

The Server Process

Figure 10.1 illustrates the basic steps that the server process goes through to establish communication. We start by creating a socket and then bind() it to a name or destination address. For local sockets, the name is a file system entry often in /tmp or /usr/tmp. For network sockets it is a *service identifier* consisting of a "dotted quad" Internet address (as in 192.168.1.11 for example) and a protocol port number. Clients use this name to access the service.

Next, the server creates a connection queue with the listen() service and then waits for client connection requests with the accept() service. When a connection request is received successfully, accept() returns a new socket, which is then used for this connection's data transfer. The server now transfers data using standard read() and write() calls that use the

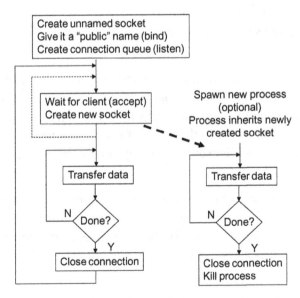

Figure 10.1
Network Server.

socket descriptor in the same manner as a file descriptor. When the transaction is complete the newly created socket is closed.

The server may very well spawn a new process or thread to service the connection while it goes back and waits for additional client requests. This allows a server to serve multiple clients simultaneously. Each client request spawns a new process/thread with its own socket. If you think about it, that's how a web server operates.

The Client Process

Figure 10.2 shows the client side of the transaction. The client begins by creating a socket and naming it to match the server's publicly advertised name. Next, it attempts to `connect()` to the server. If the connection request succeeds, the client proceeds to transfer data using `read()` and `write()` calls with the socket descriptor. When the transaction is complete, the client closes the socket.

If the server spawned a new process to serve this client, that process should go away when the client closes the connection.

Socket Attributes

The socket system call creates a socket and returns a descriptor for later use in accessing the socket.

```
#include <sys/socket.h>
int socket (int domain, int type, int protocol);
```

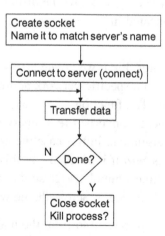

Figure 10.2
Network Client.

A socket is characterized by three attributes that determine how communication takes place. The *domain* specifies the communication medium. The most commonly used domains are PF_UNIX for local file system sockets and PF_INET for Internet connections. The "PF" here stands for Protocol Family.

The domain determines the format of the socket name or address. For PF_INET, the address is AF_INET and is in the form of a dotted quad. Here "AF" stands for Address Family. Generally, there is a one-to-one correspondence between AF_ values and PF_ values. A network computer may support many different network services. A specific service is identified by a "port number." Established network services like ftp, http, and so on have defined port numbers below 1024. Local services may use port numbers above 1023.

Some domains, PF_INET for example, offer alternate communication mechanisms. SOCK_STREAM is a sequenced, reliable, connection-based, two-way byte stream. This is the basis for TCP and is the default for PF_INET domain sockets. SOCK_DGRAM is a *datagram* service. It is used to send relatively small messages with no guarantee that they will be delivered or that they won't be reordered by the network. This is the basis of UDP. SOCK_RAW allows a process to access the IP directly. This can be useful for implementing new protocols directly in User Space.

The protocol is usually determined by the socket domain and you don't have a choice. So the protocol argument is usually zero.

A Simple Example

The Server

Time to create another Eclipse project. You know the drill. Call it "network." It's a makefile project and this one is located in home/src/network. Import the project settings with the include paths.

Open the file netserve.c in the editor. First, we create a server_socket that uses streams. Next, we need to bind this socket to a specific network address. That requires filling in a sockaddr_in structure, server_addr. The function inet_aton() takes a string containing a network address as its first argument, converts it to a binary number, and stores it in the location specified by the second argument, in this case the appropriate field of server_addr. Oddly enough, inet_aton() returns *zero* if it succeeds. In this example, the network address is passed in through the compile-time symbol SERVER so that we can build the server to run either locally through the loopback device or across the network.

The port number is 16 bits and is also passed in from the makefile through the compile-time symbol PORT. The function htons() is one of a small family of functions, macros actually, which solves the problem of transferring binary data between computer architectures with different byte-ordering policies. The Internet has established a standard "network byte order," which happens to be Big Endian. All binary data are expected to be in network byte order when it reaches the network. htons() translates a short (16 bit) integer from "host byte order," whatever that happens to be, to network byte order. There is a companion function, ntohs(), that translates back from network byte order to host order. Then there is a corresponding pair of functions that do the same translations on long (32 bit) integers.[2]

Now we *bind* server_socket to server_addr with the bind() function. Finally, we create a queue for incoming connection requests with the listen() function. A queue length of one should be sufficient in this case because there's only one client that will be connecting to this server.

Now, we're ready to *accept* connection requests. The arguments to accept() are:

- The socket descriptor.
- A pointer to a sockaddr structure that accept() will fill in.
- A pointer to an integer that currently holds the length of the structure in argument 2. accept() will modify this if the length of the client's address structure is shorter.

accept() blocks until a connection request arrives. The return value is a socket descriptor to be used for data transfers to/from this client. In this example, the server simply echoes back text strings received from the client until the incoming string begins with "q".

The Client

Now look at netclient.c. netclient determines at runtime whether it is connecting to a server locally or across the network. We start by creating a socket and an address structure in the same manner as in the server. Then, we *connect* to the server by calling connect(). The arguments are:

- The socket descriptor.
- A pointer to the sockaddr structure containing the address of the server we want to connect to.
- The length of the sockaddr structure.

When connect() returns we're ready to transfer data. The client prompts for a text string, writes this string to the socket, and waits to read a response. The process terminates when the first character of the input string is "q".

---

[2] Try to guess the names of the long functions.

Try it Out

We'll need to create make targets for the server and client. Take a look at the makefile to see how these targets are specified. This project includes a number of make targets. You can create them all now or wait until they're needed. Note incidentally that there's no all target "at all." Make both the client and the local server. Set up the host debug configuration to debug the server and start a debug session. Step through the code up to and through the call to accept(), which will block until the client initiates a connection.

In a shell window, cd to the network/ directory and execute ./netclient. Type in a few strings and watch what happens. To terminate both processes, enter a string that begins with "q" ("quit" for example).

Next we'll want to run netserve on the target with netclient running on the host. Delete netserve.o and build the make target for the remote server.

In the terminal window connected to the target (the one running minicom), cd to the network/ directory and execute ./netserve. Back in the original window execute ./netclient remote.

A Remote Thermostat

Moving on to a more practical example, our thermostat may very well end up in a distributed industrial environment where the current temperature must be reported to a remote monitoring station and setpoint and limit need to be remotely settable. Naturally, we'll want to do that over a network.

Copy monitor.c and thermostat.c from the Posix/ directory to network/ and do a Refresh on the network project to bring those files into Eclipse. Actually, thermostat.c doesn't require any changes. Here are the modifications required in monitor.c:

1. Isolate the process of setting up the net server and establishing a connection in a separate function called createServer(), which consists roughly of lines 18–55 from netserve.c. The reason for doing it this way will become apparent soon. createServer() takes no arguments and returns an int that is the socket number returned by accept(). It should return a −1 if anything fails.

2. Call createServer() from monitor() just before the while (1) loop and check the return value for error.

3. Replace fgets() with read() on the socket returned by createServer().

4. Add a new command to the switch statement, "?", that returns the current temperature over the network. Actually, we might want to query any of the thermostat's parameters. Think about how you might extend the command protocol to accomplish that.

5. The `netclient` expects a response string for every string it sends to the server. So return "OK" for every valid parameter change command and return "???" for an invalid command.

Create make targets for both the simulation and target versions of the thermostat. Build the simulation version and point the host debug configuration at `network/thermostat_s`. You'll need two shell windows to exercise the thermostat: one for the `devices` program and the other for `netclient`.

When you're satisfied with the simulation, build the target version and try it there. Remember to delete `thermostat.o` and `monitor.o`[3] before rebuilding. Also remember to execute `netclient remote` to get the client talking to the server on the target.

Multiple Monitor Threads

As the net thermostat is currently built, only one client can be connected to it at a time. It's quite likely that in a real application, multiple clients might want to connect to the thermostat simultaneously. Remember from our discussion of sockets that the server process may choose to spawn a new process or thread in response to a connection request so it can immediately go back and listen for more requests.

In our case, we can create a new monitor thread. That's the idea behind the `createServer()` function. It can be turned into a server thread whose job is to spawn a new monitor thread when a connection request comes along. So give it a try.

Here, in broad terms, are the steps required to convert the net thermostat to service multiple clients:

1. Copy `monitor.c` to `multimon.c` and work in that file.
2. Recast `createServer()` as a thread function.
3. In the `createThread()` function, change the arguments of the `pthread_create()` call to create the server thread instead of the monitor thread.
4. Go down to the last four lines in `createServer()`, following the comment "Accept a connection." Bracket these lines with an infinite while loop.
5. Following the `printf()` statement, add a call to `pthread_create()` to create a monitor thread.
6. Add a `case 'q':` to the `switch` statement in the monitor thread. "q" says the client is terminating (quitting) the session. This means the monitor thread should exit.

Now, here are the tricky parts. The `client_socket` will have to be passed to the newly created monitor thread. Where does the thread object come from for each invocation of `pthread_create()`? A simple way might be to estimate the maximum number of clients that

---

[3] There should be a way to do this automatically in the makefile, but I haven't figured it out yet.

would ever want to access the thermostat and then create an array of pthread_ts that big. The other, more general, approach is to use dynamic memory allocation, the malloc() function.

But the biggest problem is, how do you recover a pthread_t object when a monitor thread terminates? The monitor thread itself can't return it because the object is still in scope, that is, still being used. If we ignore the problem, the thermostat will eventually become unresponsive either because all elements of the pthread_t array have been used, or we run out of memory because nothing is being returned to the dynamic memory pool.

Here are some thoughts. Associate a flag with each pthread_t object whether it is allocated on a fixed array or dynamically allocated. The flag, which is part of a data structure that includes the pthread_t, indicates whether its associated pthread_t is free, in use, or "pending," that is, able to be freed. The server thread sets this flag to the "in use" state when creating the monitor. The structure should probably also include the socket number. It is this "meta" pthread_t object that gets allocated and passed to the monitor thread.

Now the question is, when can the flag be set to the "free" state? Again, the monitor can't do it because the pthread_t object is still in use until the thread actually exits. Here's where the third state, PENDING, comes in.

The monitor sets the flag to PENDING just before it terminates. Then we create a separate thread, call it "resource" if you will, that does the following:

1. Wakes up periodically and checks for monitor threads that have set their flags to PENDING.
2. Joins the thread.
3. Marks the meta pthread object free.
4. Goes back to sleep.

Remember that when one thread joins another, the "joiner" (in this case resource) is blocked until the "joinee" (monitor) terminates. So when resource continues from the pthread_join() call, the just-terminated monitor's pthread_t object is guaranteed to be free. The semantics of pthread_join() are:

```
int pthread_join (pthread_t thread, void **ret_value)
```

It's quite likely that by the time the resource thread wakes up, the monitor thread has already exited, in which case pthread_join() returns immediately with the error code ESRCH meaning that *thread* doesn't exist.

The createThread() function now needs to create the server thread and the resource thread. Likewise, terminateThread() needs to cancel and join both of those. And of course, any monitor threads that are running also need to be cancelled. That is, when the resource thread terminates, it needs to go through the list of meta pthreads checking for running monitors. Ah, but the resource thread doesn't know it's being cancelled.

Here's a good use for a cleanup handler. Write a cleanup handler that runs through the list of meta pthreads looking for any that are in the IN USE state. Cancel and join the corresponding thread. Call `pthread_cleanup_push()` just before the main loop in the resource thread. Add a call to `pthread_cleanup_pop()` just after the loop. It will never be called, but it closes the block opened by `pthread_cleanup_push()`.

If you don't feel up to writing this yourself, you may have already noticed that there is an implementation of `multimon.c` in the `home/.working/` directory.

There's a separate makefile called `Makefile.multi` to build the multiple client versions of `netthermo`. The build command for the Eclipse target is:

```
make —f Makefile.multi
```

Add `SERVER = REMOTE` to build the target version.

So how do you test multiple monitor threads? Simple. Just create multiple shell windows and start `netclient` in each one. When you run thermostat under the debugger, you'll note that a new thread is created each time `netclient` is started.

Embedded Web Servers

While programming directly at the sockets level is a good introduction to networking, and a good background to have in any case, most real-world network communication is done using higher level application protocols. HTTP, the protocol of the World Wide Web, has become the de facto standard for transferring information across networks. No big surprise there. After all, virtually every desktop computer in the world has a web browser. Information rendered in HTTP is accessible to any of these computers with no additional software.

Background on HTTP

HTTP is a fairly simple synchronous request/response ASCII protocol over TCP/IP as illustrated in Figure 10.3. The client sends a request message consisting of a header and, possibly, a body separated by a blank line. The header includes what the client wants along with some optional information about its capabilities. Each of the protocol elements shown in Figure 10.3 is a line of text terminated by CR/LF. The single blank line at the end of the header tells the server to proceed.

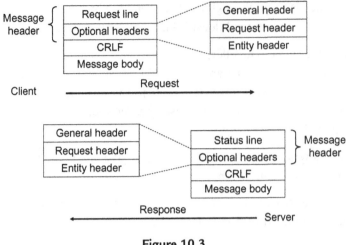

Figure 10.3
HTTP.

```
GET / HTTP/1.1
Host: 127.0.0.1
User-Agent: Mozilla/5.0 (X11; U; Linux i586; en-US; rv:1.2.1) Gecko/20030225
Accept: text/xml,application/xml,application/xhtml+xml,text/html;q=0.9,text/plain;q=0.8
Accept-Language: en-us, en;q=0.50
Accept-Encoding: gzip, deflate, compress;q=0.9
Accept-Charset: ISO-8859-1, utf-8;q=0.66, *;q=0.66
Keep-Alive: 300
Connection: keep-alive
<blank line>
```

Listing 10.1

A typical request packet is shown in Listing 10.1. The first line starts with a *method token*, in this case GET, telling the server what "method" to use in fulfilling this request. This is followed by the "resource" that the method acts on, in this case a file name. The server replaces the "/" with the default file index.html. The remainder of the first line says the client supports HTTP version 1.1.

The Host: header specifies to whom the request is directed, while the User-Agent: header identifies who the request is from. Next come several headers specifying what sorts of things this client understands in terms of media types, language, encoding, and character sets. The Accept: header line is actually much longer than shown here.

The Keep-Alive: and Connection: headers are artifacts of HTTP version 1.0 and specify whether the connection is "persistent," or is closed after a single request/response interaction. In version 1.1, persistent connections are the default. This example is just a small subset of the headers and parameters available. For our fairly simple embedded server, we can in fact ignore most of them.

A Web-Enabled Thermostat

To make data available via HTTP you need a web server. Creating a web server for an embedded device is not nearly as hard as you might think. That's because all the rendering, the hard part, is done by the client, the web browser. By and large, all the server has to do is to deliver files.

The `network/` directory contains a simple example called `webserve.c`. Open it in Eclipse. It starts out quite similar to `netserve.c` except that it listens on port 80, the one assigned to HTTP. Once the connection is established, the server reads a request message from the client and acts on it. For our rather limited purposes, we're only going to handle two methods: POST and GET.

In the case of GET, the function `doGETmethod()` near line 183 opens the specified file and determines its content type. In "real" web servers like Apache and HTML files are kept in a specific directory. In this case, it just seems easier to put the files in the same directory as the program, so `doGETmethod()` strips off the leading "/" if present to make the path relative to the current directory.

If everything is okay, we call `responseHeader()` to send the success response. The response header indicates the content type of the file and also tells the server that we want to close the connection, even if the client asked to keep the connection alive. Finally, we send the file itself. If it's an HTML file we need to parse it looking for dynamic content tags.

Dynamic Web Content

Just serving up static HTML web pages isn't particularly interesting, or even useful, in embedded applications. Usually, the device needs to report some information and we may want to exercise some degree of control over it. There are a number of ways to incorporate dynamic content into HTML, but, since this isn't a book on web programming, we're going to take a "quick and dirty" approach.

This is a good time to do a quick review of HTML. Take a look at the file `index.html` in the `network/` directory.

A nice feature of HTML is that it's easily extensible. You can invent your own tags. Of course, any tag the client browser doesn't understand it will simply ignore. So if we invent a tag, it has to be interpreted by the server before sending the file out. We'll invent a tag called `<DATA>` that looks like this:

```
<DATA data_function>
index.html has one data tag in it.
```

The server scans the HTML text looking for a <DATA> tag. data_function is a function that returns a string. The server replaces the <DATA> tag with the returned string. Open webvars.c. Near the top is a table with two entries each consisting of a text string and a function name. Just below that is the function cur_temp(), which returns the current temperature of our thermostat as an ASCII string.

Now go back to webserve.c and find the function parseHTML() around line 124. It scans the input file for the <DATA> tag. When one is found, it writes everything up to the tag out to the socket. Then it calls web_var() with the name of the associated data function. web_var(), in webvars.c, looks up and invokes the data function and returns its string value. The return value from web_var() is then written out and the scan continues.

Needless to say, a data function can be anything we want it to be. This particular example happens to return the current temperature. Incidentally, there's a subtle bug in parseHTML(). See what you can do about it.[4]

Forms and the POST Method

The <DATA> tag is how we send dynamic data from the server to the client. HTML includes a simple mechanism for sending data from the client to the server. You've no doubt used it many times while web surfing. It's the <FORM> tag. The one in our sample index.html file looks like this:

```
<FORM ACTION = "control" METHOD = "POST">
    <INPUT TYPE = TEXT NAME = "Setpoint" SIZE = 10>Setpoint:
    <INPUT TYPE = TEXT NAME = "Limit" SIZE = 10>Limit:
    <INPUT TYPE = TEXT NAME = "Deadband" SIZE = 10>Deadband:
    <INPUT TYPE = "submit" NAME = "Go">
</FORM>
```

This tells the browser to use the POST method to execute an ACTION named "control" to send three text variables when the user presses a "Go" button. Normally, the ACTION is a CGI script. That's beyond the scope of this discussion so we'll just implement a function for "control."

---

[4] Hint: What happens if a <DATA> tag spills over a buffer boundary? An expedient way around this is to read and write one character at a time. But that would incur a significant performance hit because each call to read () and write() causes a transition to kernel space with the attendant overhead. So it's better if we can read and write in larger chunks.

Have a look at the doPOSTmethod() function around line 165. It retrieves the next token in the header, which is the name of the ACTION function. Then it calls web_var(), the same function we used for the dynamic <DATA> tag. In this case, we need to pass in another parameter, a pointer to the message body because that's where the data strings are.

In the <DATA> tag case, a successful return from web_var() yields a string pointer. In the POST method case, there's no need for a string pointer, but web_var() still needs to return a non-zero value to indicate success.

The body of an HTTP POST message contains the values returned by the client browser in the form:

<name1> = <value1>&<name2> = <value2>&<name3> = <value3>&Go =

The function parseVariable() in webvars.c searches the message body for a variable name and returns the corresponding value string.

Build and Try it

Create the necessary make targets for the web thermostat and try it out. Run it under the debugger on your workstation so you can watch what happens as a web browser accesses it. You must be running as root in order to bind to the HTTP port 80.

In your favorite web browser enter

http://127.0.0.1

into the destination window. You should see something like Figure 10.4 once the file index.html is fully processed.

When running the web thermostat on the target board, enter http://192.168.1.50 into your web browser.

Once you have that running on both the workstation and target, here's an "extra credit" assignment. There are three small GIF files in the network/ directory that are circles—one red, one green, and one open. These can be used to represent the states of the cooler and alarm. So try adding a couple of <DATA > tags that will put out the appropriate circle image based on the heater and alarm states.

Of course, the other thing we would want to do is dynamically update the temperature reading. Dynamic web content gets into java applets and other issues that are beyond the scope of this book, but are definitely worth pursuing. Google the phrase "embedded web server" to find a veritable plethora of commercial products and Open Source projects that expand on what we've done here.

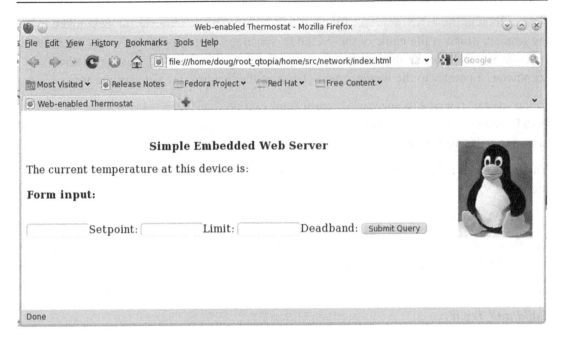

Figure 10.4
Web Thermostat.

A "Real" Web Server—Boa

While `webserve.c` is a useful introduction to HTTP processing, it is rather limited. Fortunately, the open source world includes a number of practical web servers that are suitable for embedded applications. The Resources section includes a link to a Wikipedia article that provides a very useful comparison of "light weight" web servers. Many of these are specifically intended for embedded use while others were developed as a proof of concept to see just how small one could make a web server.

The boa web server is a good example of an Open Source embedded server and it happens to be in our target file system. Boa is a single-tasking HTTP server, meaning it doesn't fork a copy of itself, or spawn a thread to handle each incoming connection, but rather internally multiplexes the connections. It first appeared publicly around 1995. The latest release is dated 2005.

One of the last things the `rcS` script does is start boa by invoking the `httpd` script in `/etc/rc.d/init.d`. But remember that we commented out the starting of `httpd` earlier in Chapter 5. So edit /etc/init.d/rcs to uncomment lines 44−47. Save the file and reboot the target board. `boa` is now running in the background waiting for a connection request to port 80.

The file `/etc/boa/boa.conf` provides configuration information for the program. Among other things, it specifies where the top level HTML files will be found. In this case it's `/www`. When you open a web page, you normally don't specify a file name. The default is usually `index.html`. This is also specified in `boa.conf`. Note that the `/www` directory is owned, and only accessible, by root.

So open a browser on your workstation and point it to http://192.168.1.50. You'll see a promotion for FriendlyARM, primarily in Chinese. There are a couple of other local pages you can view. There's also a network link that you won't be able to get to unless your target board is connected to the Internet.

Embedded E-mail

To wrap up our tour of embedded networking, let's take a quick look at how we might use e-mail in an embedded context. Suppose our thermostat maintains a log of temperatures and tracks how often the cooler comes on and for how long. We might want to periodically send that log someplace for review and analysis.

The protocol for sending e-mail is called SMTP for Simple Mail Transfer Protocol. Like HTTP, it is a fairly simple ASCII-based client−server mechanism. SMTP uses a full duplex stream socket where the client sends text commands and the server replies with responses containing numeric status codes. It is a "lock step" protocol in which every command must get a response.

Listing 10.2 is a typical dialog with an SMTP server. The client entries are bolded and extra blank lines have been added to improve readability. When the connection is first established, the server introduces itself telling us a little about it. The client in turn introduces itself with the `HELO` command, specifying a domain name. Note that each reply from the server begins with a three-digit status code. This allows for straightforward automatic handling of a message dialog, yet the accompanying text supports human interpretation. Commands and replies are not case sensitive.

Next, we specify the sender (`FROM`) and recipient (`RCPT`) of the message. There can be multiple recipients. The SMTP spec calls for e-mail addresses to be enclosed in angle brackets, at least that's what the examples show. Yet my SMTP server seems quite happy without them. The body of the message is introduced with the `DATA` command. The server responds to the `DATA` command and then doesn't respond again until the client sends the termination sequence, which is a "." on a line by itself. At this point, we could send another message or, as in this case, `QUIT`. The server closes the connection and so should the client.

```
Connection established to mail.cybermesa.com
220 smtp-out.cybermesa.com ESMTP Postfix

helo mydomain.com
250 smtp-out.cybermesa.com

mail from: me@mydomain.com
250 2.1.0 Ok

rcpt to: doug@intellimetrix.us
250 2.1.5 Ok

data
354 End data with <CR><LF>.<CR><LF>

Subject:  This is a test
This is a test of interacting with a SMTP server.
Doug
.
250 2.0.0 Ok: queued as 310831FBDC

quit
221 2.0.0 Bye
```

Listing 10.2

The program `maildialog` in the `network/` directory lets you interact with an SMTP server. That's where Listing 10.2 came from. Have a look at `maildialog.c`. The program accepts one runtime argument, the name of an SMTP mail server that can be either a dotted quad or a fully qualified domain name. Note the function `gethostbyname()`. The argument to this function is a host identifier, either a dotted quad represented as a string, or a fully qualified domain name. In the latter case, `gethostbyname()` goes out to the network to find the address of the specified domain name.

After opening a socket to the server, the program drops into a loop where it:

1. Prints the previous server response.
2. Checks for a couple of key words in the previous command that alter the loop's behavior:
   - For the `DATA` command, drop into another loop that sends text until it sees a line that just has ".".
   - For the `QUIT` command, break out and close the connection.
3. Gets another command from the user.

Build and try it with your own e-mail server.

```
make maildialog
```

```
maildialog <your_server_name>
```

The file `sendmail.c` implements a simple client that can be called from an application to send an e-mail. The mail server and local domain names are specified in string pointers that must be declared by the calling application. The message itself is embodied in a data structure. The application simply fills in the data structure and calls `sendMail()`.

Note the `do {...} while (0)` construct. This is a way of getting out of the message transaction if something goes wrong without using the dreaded `goto`. In case of an error, we simply break out of the `while` loop.

Try it out with the test application `mailclient` first.[5] Then, create a new thread for the thermostat that wakes up, say, every 15 min and sends an e-mail with the current temperature.

E-mail isn't limited to just text. Using Multipurpose Internet Mail Extensions (MIME), we can send virtually any kind of binary data—pictures, audio, video, and firmware updates—as attachments. Like most IPs, SMTP requires that arbitrary binary data be translated into 7-bit ASCII before transmission. This is usually done with Base64 encoding.

The flip side of sending an e-mail, of course, is receiving it. The most popular mechanism in use today for receiving e-mail is called the Post Office Protocol, or POP currently at version 3, known as POP3. POP3 is a command/response protocol quite similar to SMTP as illustrated in Listing 10.3, which happens to show the receive side of the send transaction shown in Listing 10.2.

What role might an embedded POP3 client serve? Automating the receipt and analysis of the log sent by our networked thermostat comes to mind. This scenario isn't necessarily a classic "embedded" application, it might very well run on a workstation. Nevertheless, it could be an element in a larger embedded application.

Like SMTP, a POP3 transaction begins with introductions, which in this case includes a password. There are a few commands, among which are `LIST` and `RETR`, which return multiple line responses terminating in a line with just a period. Having retrieved a message, we can delete it from the server. And, of course, when we're all done, we `QUIT`, closing the connection to the server.

It's a relatively trivial exercise to modify `maildialog.c` to interact with a POP3 server instead of an SMTP server. So your extra credit assignment for this section is exactly that. Create a version of `maildialog`, call it `popdialog`, that interacts with a POP3 server to retrieve email. POP3 servers normally listen at port 110.

---

[5] Don't forget to change the recipient to yourself.

```
+OK POP3 server ready <12fd299a-6c1b-47b0-a1f5-b96f32b4d0f6>

user doug%intellimetrix.us
+OK User:'doug%intellimetrix.us' ok

pass ********
+OK Password ok

list
+OK 2 messages (10114 octets)

1 9511
2 603
.

retr 2
+OK 603 octets

Return-Path: <me@mydomain.com>
Received: from mx.cybermesa.com [65.19.2.50] by dpmail12.doteasy with SMTP;
   Thu, 19 Jan 2012 15:50:23 -0800
Received: from mydomain.com (static-65-19-50-206.cybermesa.com [65.19.50.206])
      by smtp-out.cybermesa.com (Pos
tfix) with SMTP id 2ABAD1FB7A
      for <doug@intellimetrix.us>; Thu, 19 Jan 2012 16:49:34 -0700 (MST)
Subject: This is a test
Message-Id: <20120119234945.2ABAD1FB7A@smtp-out.cybermesa.com>
Date: Thu, 19 Jan 2012 16:49:34 -0700 (MST)
From: me@mydomain.com

To: undisclosed-recipients:;
X-Rcpt-To: <doug@intellimetrix.us>

This is a test of interacting with a SMTP server.
Doug
.

dele 2
+OK message 2 deleted

quit
+OK POP3 server signing off
```

Listing 10.3

Turns out there are a couple of "gotchas." The POP3 protocol insists that commands be terminated by both carriage return (\r) *and* line feed (\n). Then there are a couple of problems related to searching for the termination string, ".", on a line by itself. First, the server may send multiple lines as a single string, so you can't rely on the period being the first character. You also need to search the string for "\n.". Second, to verify that the period is on the line by itself, it's not sufficient to just search for "\n.\n" because the server may insert a carriage return, "\r", character. So following the "\n." you need to look for either "\n" or "\r".

Other Application-Level Protocols

The three protocols we've looked at, HTTP, SMTP, and POP3, can be useful for certain classes of embedded problems. There are a great many other application-level protocols that can also be useful in other situations. Simple Network Management Protocol (SNMP), for example, has become the de facto standard mechanism for managing network resources. Virtually, every device attached to a network implements SNMP and your embedded devices probably should too. In an embedded scenario, SNMP can be used for simple monitoring and control.

We've pretty much wrapped up our tour of application-level programming. In Chapter 11 we dive down into Kernel space and have a look at the Linux kernel.

Resources

Comer, Douglas, *Internetworking with TCP/IP, Vols. 1, 2 and 3*, Prentice-Hall. This is the classic reference on TCP/IP. Volume 1 is currently up to its fifth edition dated 2005. Volume 2, coauthored with David L. Stevens, is at the third edition, 1998, and volume 3, also coauthored with Stevens, dates from 1997. Highly recommended if you want to understand the inner workings of the Internet.

http://en.wikipedia.org/wiki/Comparison_of_lightweight_web_servers—This article features a pair of quite useful tables comparing more than a dozen small web server implementations including boa.

http://www.boa.org/—The home page for boa.

IPs are embodied in a set of documents known as "RFCs," Request for Comment. The RFCs are now maintained by a small group called the RFC Editor. The entire collection of RFCs, spanning the 30-plus-year history of the Internet, is available from www.rfc-editor.org. In particular, HTTP is described in RFC 2616, SMTP is described in RFC 821, and POP3 is described in RFC 1081.

Jones, M. Tim, *TCP/IP Application Layer Protocols for Embedded Systems*, Charles River Media, 2002. The idea of adding a <DATA> tag to HTML came from this book. It covers a wide range of application-level network protocols that can be useful in embedded scenarios.

Linux Network Administrators' Guide, available from the Linux Documentation Project, www.tldp.org. Not just for administrators, this is a quite complete and quite readable tutorial on a wide range of networking issues.

## CHAPTER 11

Configuring and Building the Kernel

Hackito ergo sum

Anonymous

One of the neatest things about Linux is that you have the source code. You're free to do whatever you want with it. Most of us have no intention, or need, to dive in and directly hack the kernel sources. But access to the source code does mean that the kernel is highly configurable. That is, you can build a kernel that precisely matches the requirements, or limitations, of your target system.

Now again, if your role is writing applications for Linux, as we've been doing in the last few chapters, or if you're a Linux system administrator, you may never have to touch the kernel. But as an embedded systems developer, you will most certainly have to build a new kernel, probably several times, either for the workstation or the target environment. Fortunately, the process of configuring and building a kernel is fairly straightforward. My experience has been that building a new kernel is a great confidence-building exercise, especially if you're new to Linux.

The remainder of this chapter details and explains the various steps required to configure and build an executable kernel image for our target board.

Getting Started

The source code for the Linux kernel resides at www.kernel.org where you can find the code for every version of the kernel back to Linus' original release in 1991. The sources are available in both gzipped tar format (.tgz) and the slightly smaller bzipped format (.bz2).

In general, you're free to use whichever kernel version meets your needs and/or strikes your fancy. For the moment though, you'll want to get version 3.1.5 because that's what happens to be burned in the target board's flash. Navigate to `/pub/linux/kernel/v3.0` and download `linux-3.1.5.tar.bz2`. Create a directory `arm/` in `/usr/src` and unzip the archive there. First though, check to be sure `/usr` and `/src` have read and write permission for "other" enabled so that you can work with the kernel as a normal user. I chose to create the `arm/` directory to make it clear that we're building this kernel for an ARM target.

I usually open archive files with the KDE archiver, which will automatically recognize the file format and choose the appropriate decompression utility. But for files as large as a kernel source tree, I find the shell command to be faster. `cd` to `/usr/src/arm`, if you're not there already, and execute

```
tar -xjf linux-3.1.5.tar.bz2
```

After untarring the file, you'll find a new directory in `/usr/src/arm`, `linux-3.1.5/`. It's not at all unusual to have several versions of the Linux kernel and corresponding source code on your system. How do you cope with these multiple versions? There is a simple naming strategy that makes this complexity manageable.

Generally, Linux sources are installed as subdirectories of `/usr/src`. The subdirectories usually get names of the form

```
linux-<version_number>-<additional_features>.
```

Kernel Version Numbering

This gives us an excuse to digress into how kernel versions are numbered. `<version_number>` identifies the base version of the kernel as obtained from kernel.org and looks something like this: 3.1.5. The first number is the "version", in this case 3. Normally, this number increments only when truly major architectural changes are made to the kernel and its APIs. The version number was 2 from 1996 until mid-2011 when Linus arbitrarily rolled the version to 3 in honor of the 20th anniversary of Linux.

The second number, 1, is used to identify releases where the kernel APIs may differ but the differences aren't enough to justify a full version change. New releases appear about every 2−3 months.

The final number, 5 in this example, represents bug and security fixes that don't change the kernel APIs. Applications built to a specific release should, in principle, run on any security level.

`<additional_features>` is a way of identifying patches made to a stock kernel to support additional functionality. We've installed a stock 3.1.5 kernel and so the `<additional_features>` field is null. Later on we'll create a specific configuration of this kernel and add the appropriate `<additional_features>` field.

The `<additional_features>` field is also used by the kernel developers to identify developmental source trees known as *release candidates*. Starting from a stable kernel release, sufficiently stable new code is merged into the mainline to become the next release −rc1. So, for example, `3.1.5` would become `3.2-rc1`. A period of 6−10 weeks of testing and patching ensues with the `rc` number being rolled about every week. When the new tree is deemed sufficiently stable, the `−rcn` is dropped and the process starts over again.

Given that you may have multiple kernel source trees in your file system, the kernel you're currently working with is usually identified by a symbolic link in /usr/src called linux, although in this case linux will be in /usr/src/arm. Create the symbolic link with this shell command:

```
ln -s linux-3.1.5 linux
```

The Kernel Source Tree

Needless to say, the kernel encompasses a very large number of files—C sources, headers, makefiles, scripts, etc. Version 3.1.5 has 37,800 files taking up 317 MB. So not surprisingly, there's a standard directory structure to organize these files in a manageable fashion. Figure 11.1 shows the kernel source tree starting at /usr/src/arm/linux. The directories are as follows:

Documentation—Pretty much self-explanatory. This is a collection of text files describing various aspects of the kernel, problems, "gotchas," and so on. There are many subdirectories under Documentation for topics that require more extensive explanations. While the information here is generally useful, it also tends to be dated. That is, the documentation was initially written when a particular feature was added to the kernel, but it hasn't been kept up as the feature has evolved.

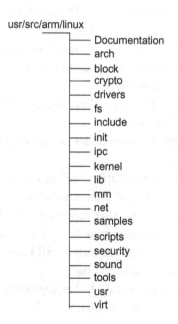

Figure 11.1
The Kernel Source Tree.

arch—All architecture-dependent code is contained in subdirectories of arch. Each architecture supported by the kernel has a directory under arch with its own subdirectory structure. The executable kernel image will end up in arch/ <architecture>/boot. An environment variable in the makefile, ARCH, points to the appropriate target architecture directory. arch/ is one of the two largest subtrees in the kernel, comprising almost 12,400 files.

block—The block layer. This is the code that optimizes transfers with block storage devices.

crypto—Code dealing with the cryptographic aspects of system security.

drivers—Device driver code. Under drivers, there are a large number of subdirectories for various devices and classes of device. This is the other major kernel subtree with close to 9,600 files.

firmware—This is a relatively new addition to the kernel tree. It contains binary firmware files for devices that load firmware at boot time. Ultimately, these files are to be moved to User Space because they are, for the most part, proprietary and mixing them with the GPLed kernel code is "problematic."

fs—File systems. Under fs, there is a set of directories for each type of file system that Linux supports.

include—Header files. The most important subdirectory of include/ is linux/ that contains headers for the kernel APIs.

init—The basic initialization code.

ipc—Code to support Unix System 5 Inter-Process Communication mechanisms such as semaphores, message passing, and shared memory.

kernel—This is the heart of the matter. Most of the basic architecture-independent kernel code that doesn't fit in any other category is here. This includes things like the scheduler and basic interrupt handling.

lib—Several utility functions that are collected into a library.

mm—Memory management functions.

net—Network support. Subdirectories under net contain code supporting various networking protocols.

samples—Another relatively new addition, this has sample code showing how to manage some of the kernel's internal data structures.

scripts—Text files and shell scripts that support the configuration and build process.

security—Offers alternative security models for the kernel.

sound—Support for sound cards and the Advanced Linux Sound Architecture (ALSA).

tools—A performance monitor for the kernel.

usr—Mostly assembly code that sets up linker segments for an initial RAM disk.

virt—A virtual environment for the kernel.

Kernel Makefile

usr/src/arm/linux contains a standard make file, Makefile, with a very large number of make targets. By default, the kernel is built for the architecture on which the makefile is running, which in the vast majority of cases is some variant of the x86. In our case, we want to cross-compile the kernel for our ARM target board.

As noted above, there is an environment variable, ARCH, that can be set to select the architecture to build for. This can be set on the command line that invokes make as in:

```
make ARCH = arm
```

Or, what I prefer to do is edit the makefile to permanently set the value of ARCH. Open the makefile in /usr/src/arm/linux with your favorite editor[1] and go to line 195, which initially reads:

```
ARCH   ? = $(SUBARCH)
```

Change it to read:

```
ARCH   ? = arm
```

The next line defines the variable CROSS_COMPILE. This identifies a cross-tool chain by specifying a prefix to be added to the tool name. Our ARM cross-tool chain is identified by the prefix arm-linux-. You can enter that here or, optionally, specify it as part of the configuration process. Save the file and exit from the editor.

Now execute make help. You'll see a very long list of make targets in several categories. Our specific interest at the moment is the category Architecture specific targets (arm). Under that, there are a large number of target names that end with "_defconfig." These make default configuration files for various ARM-based boards. When building a new kernel, it's always good practice to start with a known good configuration. In our case, that's mini2440_defconfig. So execute:

```
make mini2440_defconfig
```

There's a new file in /usr/src/arm/linux -.config.

---

[1] There's not much point in using Eclipse for configuring and building the kernel. We won't be editing any source files and the kernel has its own graphical configuration tools.

Patching the Kernel

Before we get on to actually configuring and building the kernel, there's one more hoop we have to jump through in our particular situation. This gives us an opportunity to look at the notion of "patching" Open Source code trees.

It turns out that YAFFS, the file system that's used on the target board's NAND flash is not incorporated into the mainline kernel source. This is apparently a licensing issue. Although YAFFS is released under the GPL, its copyright owner, a British company, Aleph One, retains the right to charge for a license under certain circumstances. This is enough to keep YAFFS out of the truly free Linux kernel.

Aleph One provides a gzipped tar file for YAFFS including a script that installs it in a kernel source tree. But since my objective here is to introduce the patch mechanism, I chose to execute their script against the "vanilla" (unmodified) 3.1.5 kernel source tree from kernel.org and then create a *patch*.

The mechanism for changing released source code in an orderly manner is the `patch` utility. The input to `patch` is a text file created by the `diff` utility that compares two files and reports any differences. So when an Open Source programmer wants to distribute an upgrade to released source code, he or she does a `diff` between the modified code and the original code redirecting the output of `diff` to a patch file.

In this case, I created two kernel source trees from `linux-3.1.5.tar.bz2` and renamed the first one `linux-3.1.5-orig/`. I executed Aleph One's script on the other source tree and then, from `/usr/src/arm`, I executed:

```
diff —uprN linux-3.1.5-orig linux>yaffs.patch
```

to create an appropriate patch file, `yaffs.patch`. See the `diff man` page to learn what the various options represent. Copy (move) that file from `factory_images/` to `/usr/src/arm/ linux` and execute:

```
patch —p1< yaffs.patch
```

Note that `patch` normally takes its input from `stdin`, so we have to redirect the input to the file. The `—p1` flag tells `patch` to remove one slash and everything before it from the names of files to be patched. This recognizes the usual situation that the root directory from which the patch was created is not the same as the directory in which we're applying it. Removing everything in the path before the first slash makes the directories relative.

Everything necessary to support YAFFS is now added to the kernel source tree.

This patch also installs the `mini2440_*` drivers that the Posix and later thermostat examples used along with a revised `.config` file. The default configuration leaves a lot of things turned that you don't really need. The revised.config has many of these features turned off and is in fact the configuration for the kernel in the target flash.

Configuring the Kernel—make config, menuconfig, xconfig

The process of building a kernel begins by invoking one of the three make targets that carry out the configuration process. `make config` starts a text-based script that sequentially steps you through each configuration option. For each option you have either three or four choices. The three choices are: "y" (yes), "n" (no) and "?" (ask for help). The default choice is shown in upper case.

Some options have a fourth choice, "m", which means build this feature as a loadable kernel module rather than build it into the kernel image. Kernel modules are a way of dynamically extending the kernel and are particularly useful for things like device drivers. The next chapter goes into modules in some detail.

Figure 11.2 shows an excerpt from the `make config` dialog.

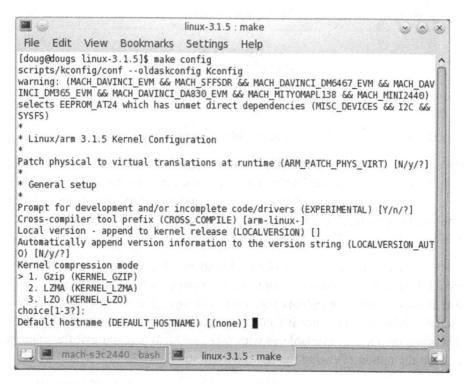

Figure 11.2
make config Dialog.

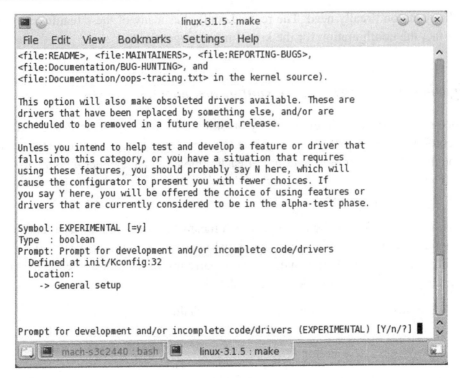

Figure 11.3
Kernel Configuration Help Text.

Most options include help text that is genuinely "helpful" (Figure 11.3).

The problem with make config is that it's just downright tedious. Typically, you'll only be changing a very few options and leaving the rest in their default state. But make config forces you to step through each and every one, and contemporary kernels have well over 2,000 options. Personally, I've never used make config and I honestly wonder why it's still in the kernel package.

make menuconfig, based on the ncurses library, brings up the pseudographical screen shown in Figure 11.4. Here the configuration options are grouped into categories and you only need to visit the categories of options you need to change. The interface is well explained and reasonably intuitive. But since it's not a true graphical program, the mouse doesn't work. The same help text is available as with make config. When you exit the main menu, you are given the option of saving the new configuration.

While most true Linux hackers seem to prefer menuconfig, my choice for overall ease of use is make xconfig. This brings up an X Windows-based menu as shown in Figure 11.5. Now

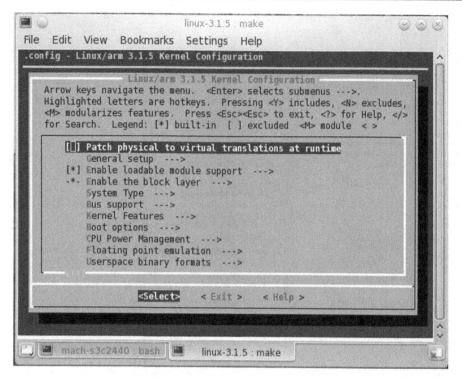

Figure 11.4
make menuconfig.

you can see all the option categories at once and navigate with the mouse. Of course, you must be running X Windows to use this option and you must have the g++, Qt graphics library, and Qt development tool packages installed.

g++ and the development tools are usually not installed by default. If they aren't, execute as root user:

```
yum install gcc-g++
yum install qt3-devel
```

As usual, Debian/Ubuntu users should replace yum with apt-get.

In its default display mode, xconfig displays a tree of configuration categories on the left. Selecting one of these brings up the set of options for that category in the upper right window. Selecting an option then displays help for that option in the lower right window. Figure 11.5 starts with the general setup menu. Most of the options are presented as check boxes. Clicking on the box alternates between checked (selected) and unchecked (not selected). These are termed *binary* options. Options that may be built as

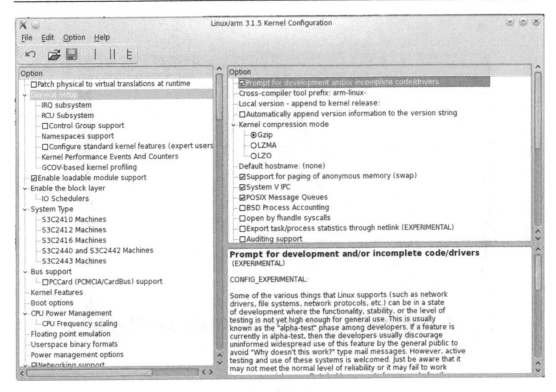

Figure 11.5
make xconfig.

kernel modules have a third value, a dot to indicate that the feature will be built as a module. These are *tristate* options. An example of that is Bus support. Click the box next to PCCard (PCMCIA/CardBus) support and it changes to a dot. Click it again and it changes to a check mark. Click it again to leave the box unchecked.

Some options have a set of allowable values other than yes or no and are represented by *radio buttons*. Select Kernel features to see a couple of examples: Memory split and Preemption Model. Some options take numeric values. An example of a numeric option is back up under General setup, kernel log buffer size. It's not immediately obvious how you change a numeric value. Double-clicking the option opens a dialog box between the two right-hand side windows.

Options that aren't available, because some other option was not selected, simply don't show up. Again, the PC Card option serves as an example. With the box unchecked, the upper right-hand side window is empty. Now click the check box and the appropriate options show up in the upper right.

Try it Out

The best way to get familiar with kernel configuration options is to fire up `xconfig` and see what's there. So. . .

```
cd /usr/src/arm/linux
make xconfig
```

After a bit of program and file building, the menu of Figure 11.5 will appear. Just browse through the submenus and their various sub-submenus to get a feel for the flexibility and richness of features in the Linux kernel. Read the help descriptions.

All of the critical configuration options should be set correctly in the revised .config file, but for additional practice, check on them. Under General setup is an option called Cross-compiler tool prefix. This is a string option that must be set to "arm-linux-." Go into Miscellaneous file systems under File systems and see that yaffs2 file system support is selected. Build it into the kernel, not as a module. You can deselect journaling Flash File System support. Be sure that NFS client support and Root file system on NFS are selected under Network File Systems.

The default configuration enables any number of options that you don't really need, at least not initially. The Mini2440 has no built-in wireless support, so Bluetooth and anything called "Wireless" can be safely removed. Native language support is another place to remove a lot of unnecessary modules.

If you do make changes, be sure to add a Local version string under General setup.

For the time being leave the configuration menu up while you read the next section.

---

make `xconfig` gives you additional flexibility in saving configurations. Instead of always saving it to the standard `.config` file, you can save it to a named file of your choice using the `File->Save As` option and later load that file using `File->Load` for further modification or to make it the new default configuration.

xconfig Options

The `xconfig` menu has several useful options in the **Option** menu. **Show Name** adds a column to the upper right panel that displays the name of each configuration variable. **Show Range** displays the range and current value of binary and tristate options. In my estimation, this is only marginally useful. Another marginally useful option is **Show Data**. For binary and tristate options it pretty much duplicates **Show Range**. It can be helpful for numeric and text options.

Show All Options is largely self-explanatory. It displays in gray all of the options that can't be selected because some condition is not satisfied. Suppose you're pretty sure some option exists, but you can't find it. Turn on **Show All Options**, then use **Edit - > Find** to search for the name. Now turn on **Show Debug Info** to see what the conditions are for that option. This feature also shows you the option name and where it's defined.

When you're finished, close the xconfig menu. You'll be asked if you want to save or discard the changes.

.config File

It's probably worth mentioning that true kernel hackers tend to prefer make `menuconfig` over `make xconfig`. The end result of the configuration process, whichever one you choose to use, is a file called `.config` containing all of the configuration variables. Listing 11.1 is an excerpt. The first thing to notice is the comment at the top:

```
# Automatically generated file; DO NOT EDIT
```

It is considered "bad form" to manually edit files that have been created by a tool. There's a high probability that you'll mess something up.

The options that are not selected remain in the file but are "commented out." The selected options get the appropriate value, "y" for features to be built into the kernel image, "m" for modules, a number or an element from a list. `.config` is included in the Makefile where it

```
#
# Automatically generated file; DO NOT EDIT.
# Linux/arm 3.1.5 Kernel Configuration
#
CONFIG_ARM=y
CONFIG_SYS_SUPPORTS_APM_EMULATION=y
CONFIG_HAVE_SCHED_CLOCK=y
CONFIG_GENERIC_GPIO=y
# CONFIG_ARCH_USES_GETTIMEOFFSET is not set
CONFIG_GENERIC_CLOCKEVENTS=y
CONFIG_GENERIC_CLOCKEVENTS_BROADCAST=y
CONFIG_KTIME_SCALAR=y
CONFIG_HAVE_PROC_CPU=y
CONFIG_STACKTRACE_SUPPORT=y
CONFIG_LOCKDEP_SUPPORT=y
CONFIG_TRACE_IRQFLAGS_SUPPORT=y
CONFIG_HARDIRQS_SW_RESEND=y
CONFIG_GENERIC_IRQ_PROBE=y
CONFIG_RWSEM_GENERIC_SPINLOCK=y
CONFIG_ARCH_HAS_CPUFREQ=y
CONFIG_ARCH_HAS_CPU_IDLE_WAIT=y
CONFIG_GENERIC_HWEIGHT=y
CONFIG_GENERIC_CALIBRATE_DELAY=y
CONFIG_NEED_DMA_MAP_STATE=y
CONFIG_VECTORS_BASE=0xffff0000
```

Listing 11.1
Excerpt of `.config`.

controls what gets built and how. Early in the build process `.config` is converted into a header file, `include/linux/autoconf.h`, so that the configuration variables can be used to control conditional compilation.

Behind the Scenes—What's Really Happening

The information in this section is not essential to the process of building a kernel and you're free to ignore it for now. But when you reach the point of developing device drivers or other kernel enhancements (or perhaps hacking the kernel itself), you will need to modify the files that control the configuration process.

All the information in the configuration menus is provided by a set of text files named `Kconfig` that are scattered throughout the source tree. These are script files written in Config Language, which looks suspiciously like a shell scripting language but isn't exactly. The main `Kconfig` file is located in `linux/arch/$(ARCH)` where `ARCH` is a variable in `Makefile` identifying the base architecture.

Go to `linux/arch/arm` and open `Kconfig`. Find the line that reads `menu "System Type"` around line 222. Compare the structure of the file with the configuration menu starting at System Type and the pattern should become fairly clear. Each configuration option is identified by the keyword `config` followed by the option's symbolic name, for example `config MMU`. Each of these gets turned into a Makefile variable, or macro, such as `CONFIG_MMU`. The Makefiles then use these variables to determine which components to include in the kernel and to pass `#define` symbols to the source code.

The option type is indicated by one of the following key words:

- `bool`—The option has two values, "y" or "n".
- `tristate`—The option has three values, "y", "n", and "m".
- `choice`—The option has one of the listed values.
- `int`—The option takes an integer value.

The type key word usually includes a prompt, which is the text displayed in the configuration menu. Alternatively, the `prompt` key word specifies the prompt. There are also "hidden" options that have no displayable prompt. At the top of `Kconfig` is a set of options that don't show up in the menu because no prompt is specified. In the configuration menu, click on `Options->Show All Options`. Now all options are displayed, including those with no prompt and those dependent on unselected options.

Other key words within a configuration entry include `help` to specify the help text and `depends`, which means that this option is only valid, or visible, if some other option that it depends on is selected. There's also `select`, which says that if this option is selected, then of necessity some other option must also be turned on.

Just above the `menu "System Type"` line is the line:

```
source init/Kconfig
```

The `source` key word is how the configuration menu is extended to incorporate additional, modular features. It's effectively the same thing as `#include` in C. You'll find `source` lines scattered throughout the main `Kconfig` file.

Config Language is actually much more extensive than this simple example would suggest. For more detail, look at `linux/Documentation/kbuild/kconfig-language.txt`.

Building the Kernel

The actual process of building the kernel varies a little depending on what you're building it for. Our primary objective is to build a kernel for the target board so I'll describe that first. Then, I'll describe the alternate process of building a new kernel for your workstation.

The first three steps below can be executed as a normal user.

> `make clean`. Deletes all intermediate files created by a previous build. This insures that *everything* gets built with the current configuration options. You'll find that virtually all Linux makefiles have a `clean` target. Strictly speaking, if you're building for the first time, there are no intermediate files to clean up so you don't have to run `make clean`.
> `make`. This is the heart of the matter. This builds the executable kernel image and all kernel modules. Not surprisingly, this takes a while. The resulting compressed kernel image is `arch/$(ARCH)/boot/bzImage`.
> `make uImage`. The u-boot boot loader on the target board requires a 64-byte header in front of any file it deals with. The `uImage` make target puts that header in front of `bzImage`, creating the file `uImage` in `arch/$(ARCH)/boot`.

The 64-byte header tells U-boot about the nature of the file data. The header conveys image properties including:

- Target OS
- Target CPU architecture
- Type of image—kernel, filesystem, stand-alone program, etc.
- Compression type
- Load Address
- Entry Point
- Image Name
- Image Timestamp.

Previously, in Chapter 4 we copied an executable called `mkimage` to the `bin/` directory of the ARM cross-tool chain. This is the utility that builds the header. It is invoked by the `uImage`

Makefile target. `mkimage` is actually part of the U-boot distribution that we'll get to in Chapter 14.

The following step requires root user privileges.

`make modules_install`

`INSTALL_MOD_PATH` = /home/<your_user_name>/root_qtopia. Copies the modules to `$(INSTALL_MOD_PATH)/lib/modules/<kernel_version>` where `<kernel_version>` is the string identifying the specific kernel you are building. Note that when building for the target board you must explicitly specify the root file system. Otherwise, the modules will go into `/lib/modules` of your workstation's file system. Strictly speaking, this step isn't necessary in this case because none of the modules are required to boot the kernel.

That's it. Note incidentally that the build process is recursive. Every subdirectory in the kernel source tree has its own Makefile dealing with the source files in that directory. The top-level Makefile recursively invokes all of the sub-Makefiles.

The process for building a kernel for your workstation differs in a couple of details. You don't do the `make uImage` step. Do `make modules_install` without the `INSTALL_MOD_PATH` argument and then execute this step, also as root:

`make install`. This does several things. It copies the kernel executable, called `vmlinuz` for x86 builds, to `/boot` along with `System.map`, the linker map file. It adds a new entry to `/boot/grub/grub.conf` so the GRUB can offer the new kernel as a boot option. Finally, it creates an *initial ramdisk*, initrd, in `/boot`.

Workstation Digression

Before we move on to loading and testing our new kernel, the last two items mentioned above deserve a passing explanation. Most workstation installations these days incorporate a boot loader called GRUB to select the specific kernel or alternate OS to boot. There's a very good reason for having the ability to boot multiple kernel images. Suppose you build a new kernel and it fails to boot properly. You can always go back and boot a known working image and then try to figure out what went wrong in your new one. If you're curious, have a look at the file `/boot/grub/grub.conf` on your workstation.

Most Linux kernels are set up to use `initrd`, which is short for initial ramdisk. An initial ramdisk is a very small Linux file system loaded into RAM by the boot loader and mounted as the kernel boots, before the main root file system is mounted. The usual reason for using initrd is that some kernel modules need to be loaded before mounting the root partition. Usually, these modules are required to support the file system used by the root partition, ext3 for example, or perhaps the controller that the hard drive is attached to, such as SCSI or RAID.

Booting the New Kernel

Now that we have a new kernel, how do we test it? We could of course load the new image into flash, either in addition to, or in place of, the kernel image presently there.

The other alternative is to boot the new image over the network using TFTP. This is particularly advantageous in a development environment because we can quickly test a new kernel image without the time-consuming process of burning flash.

Move `arch/arm/boot/uImage` to `/var/lib/tftpboot`, or wherever else you chose as the tftp directory. Boot the target board into U-boot and execute the following commands:

```
tftpboot 32000000 uImage
bootm 32000000
```

`tftpboot` downloads `uImage` to location 0x32000000 in RAM. Then `bootm` interprets the U-boot header at 0x32000000, copies the image to the correct location in memory, uncompressing if necessary, and transfers control to it. After the kernel boots, execute the command `uname -a` and note that the kernel image has a timestamp corresponding to when you built it.

Congratulations! You can pat yourself on the back. You're now a Linux hacker.

Note by the way that U-boot always interprets numbers as hex. Also, there's a u-boot environment variable, `boot_tftp`, that includes the two commands listed above. Run that to tftp and boot the kernel.

In the next chapter, we'll dive down into kernel space and look at the resources for writing Linux device drivers.

Resources

For more details on the process of configuring and building a kernel, look at the files in `/usr/src/arm/linux/Documentation/kbuild`. Additionally, the following HOW-TOs at www.tldp.org may be of interest:

`Config-HOWTO`   This HOWTO is primarily concerned with how you configure your system once it's built.

`Kernel-HOWTO`   Provides additional information on the topics covered in this chapter.

www.linuxfromscratch.org/—This is a project that provides you with step-by-step instructions for building your own custom Linux system, entirely from source code. There's a subproject that deals with cross building Linux for embedded environments.

www.yaffs.net—The home page for YAFFS.

## CHAPTER 12

Kernel Modules and Device Drivers

If you think I'm a bad driver, you should see me putt.

Snoopy

Our objective in this chapter is to introduce the basics of Linux device drivers. We'll begin with the concept of installable kernel modules. Then we'll look at the basic device driver APIs in the context of a simple GPIO driver for the Mini2440. We'll also look at how interrupts are handled. Finally, we'll see how to integrate a new device driver into the kernel source tree.

Of necessity, this is a cursory overview of a very complex topic. The resources section offers additional reading.

Kernel Modules

Installable kernel modules offer a very useful way to extend the functionality of the basic Linux kernel and add new features without having to rebuild the kernel. A key feature is that modules may be dynamically loaded when their functionality is required and subsequently unloaded when no longer needed. You might think of modules as the kernel equivalent of User Space processes. Modules are particularly useful for things like device drivers and file systems.

Given the range and diversity of hardware that Linux supports, it would be impractical to build a kernel image that included all of the possible device drivers. Instead the kernel includes only drivers for boot devices and other common hardware such as serial and parallel ports, IDE and SCSI drives, and so on. Other devices are supported as loadable modules and only the modules needed in a given system are actually loaded.

There's usually not a good technical reason for using loadable modules in a production-embedded environment because we know in advance exactly what hardware the system must support and so we simply build that support into the kernel image. Nevertheless, we'll see later on that there might be a "business" reason for using modules. Modules are still useful when testing a new driver. You don't need to build a new kernel image every time you find and fix a problem in your driver code. Just load it as a module and try it.

Keep in mind that modules execute in Kernel Space at Privilege Level 0 and thus are capable of bringing down the entire system.

Linux for Embedded and Real-time Applications.
© 2013 Elsevier Inc. All rights reserved.

DOI: http://dx.doi.org/10.1016/B978-0-12-415996-9.00012-5

A Module Example

Create a new Eclipse Makefile project in the `home/src/hello/` directory. Open the file `hello.c`. Note, first of all the two include files required for a kernel module. This is a trivial example of a loadable kernel module. It contains two very simple functions; `hello_init()` and `hello_exit()`. The last two lines:

```
module_init(hello_init);
module_exit(hello_exit);
```

identify these functions to the module loader and unloader utilities. Every module must include an init function and an exit function. The function specified to the `module_init()` macro, in this case `hello_init()`, is called by `insmod`, the shell command that installs a module. The function specified to `module_exit()`, `hello_exit()`, is called by `rmmod`, the command that removes a module.

You probably noticed that Eclipse reported a number of semantic and syntax errors and code analysis problems. These are bogus as you'll find that the project does in fact build correctly. The Eclipse CDT folks concede that static code analysis is "far from perfect" and does generate false errors. If you find these messages annoying, you can turn off static analysis by opening the project properties dialog and going to C/C++ General > Code Analysis. Select Use project settings and uncheck Potential Programming Problems and Syntax and Semantic Errors. Of course, you'll also need to add the appropriate include path, which is `<absolute_path_to_kernel_source>/include`.

In this example, both functions simply print a message on the console using `printk()`, the kernel equivalent of `printf()`. C library functions like `printf()` are intended to run from user space making use of OS features like redirection. These facilities aren't available to kernel code. Rather than writing directly to the console, `printk()` writes to a circular buffer and then wakes up the `klogd` daemon to write the message to the system log and perhaps print it on the console.

Note the `KERN_ALERT` at the beginning of the `printk()` format strings. This is the symbolic representation of a *loglevel* that determines whether or not the message appears on the console. Loglevels range from 0 to 7 with lower numbers having higher priority. If the loglevel is numerically less than the kernel integer variable `console_loglevel`, then the message appears on the console. In any case, `printk` messages always show up in the file `/var/log/messages` regardless of the loglevel.

The `hello` example also shows how to specify module parameters, local variables whose values can be entered on the command line that loads the module. This is done with the `module_param()` macro. The first argument to `module_param()` is a variable name, the second is the variable type, and the third is a "permission flag" that controls access to the representation of the module parameter in *sysfs*, a new feature of the 2.6 kernel that offers cleaner access to driver internals than the /proc filesystem. A safe value for now is S_IRUGO meaning the value is read-only.

Variable types can be charp—pointer to a character string, or int, short, and long—various size integers or their unsigned equivalent with "u" prepended.

Build the project and then try it out. Note that even though the project resides in the target's root file system, it is intended to run on the workstation. In a shell window, as root user in the module/ directory, enter the command

```
insmod hello.ko my_string = "name" my_int = 47
```

If you initially logged on under your normal user name and used su to become root, you'll have to preface all of the module commands with /sbin/ because /sbin is not normally part of the path for a normal user. printk messages don't appear in shell windows running under X windows regardless of the loglevel. That's because X windows is running "virtual" terminals. You can see what printk did with the command:

```
tail /var/log/messages
```

again as root user. This prints out the last few lines of the messages log file. A useful variation on this command is:

```
tail —f /var/log/messages
```

The "f" means "follow." This version continues running and outputs new text sent to messages. The command can also be expressed as tailf. I usually keep a second shell window open to watch the messages file.

Modules are kernel version-specific. That is, insmod will only load modules that have been compiled for the currently running kernel version. That's because kernel APIs can and do change over time. A module that was compiled for, say, version 3.1.5 may very well crash if it's loaded into version 3.4.4 because the argument list for a kernel function changed. insmod does offer an option to force loading of modules compiled for other versions (-f) and in many cases it may work.

Try the command lsmod. This lists the currently loaded modules in the reverse order in which they were loaded. hello should be the first entry. lsmod also gives a "usage count" for each module and shows what modules depend on other modules. The same information is available in the file /proc/modules.

Now execute the command rmmod hello. You should see the message printed by hello_exit(). Finally execute lsmod again to see that hello is no longer listed. Note incidentally that rmmod does not require the ".ko" extension. Once loaded, modules are identified by their base name only.

A module may, and usually does, contain references to external symbols such as printk. How do these external references get resolved? insmod resolves them against the kernel's symbol table, which is loaded in memory as part of the kernel boot process. Furthermore, any exported symbols defined in the module are added to the kernel symbol table and are available for use by subsequently loaded modules. So the only external symbols a module can reference are those built into the kernel image or previously loaded modules. The kernel symbol table is available in /proc/ksyms.

"Tainting" the Kernel

You should have seen another message when you installed hello, just before the message from hello_init():

```
hello: module license 'unspecified' taints kernel
```

What the heck does that mean? Apparently, the kernel maintainers were getting tired of trying to cope with bug reports involving kernel modules for which no source was available, that is, modules not released under an Open Source license such as GPL. Their solution to the problem was to invent a MODULE_LICENSE() macro whereby you can declare that a module is indeed Open Source. The format is:

```
MODULE_LICENSE ("<approved string>")
```

where <approved_string> is one of the ASCII text strings found in linux/include/linux/module.h. Among these, not surprisingly, is the "GPL." If you distribute your module in accordance with the terms of an Open Source license such as GPL, then you are permitted to include the corresponding MODULE_LICENSE() invocation in your code and loading your module will not produce any complaints.

If you install a module that produces the above warning and the system subsequently crashes, the crash documentation (core dump) will reveal the fact that a non-Open Source module was loaded. Your kernel has been "tainted" by code that no one has access to. If you submit the crash documentation to the kernel maintainers it will be ignored.[1]

---

[1] What's not clear to me is how the tainted kernel provision is enforced. An unscrupulous device vendor could very easily include a MODULE_LICENSE() entry in his driver code but still not release the source. What happens when that module causes a kernel fault? I suspect the Open Source community is relying on public approbation to "out" vendors who don't play by the rules. What else is there?

Add the following line to `hello.c` just below the two `module_param()` statements:

```
MODULE_LICENSE("GPL");
```

Remake hello and verify that installing the new version does not generate the warning. Nevertheless, your kernel remains tainted. The tainted flag remains set until you reboot the system. The tainted flag is visible at `/proc/sys/kernel/tainted`. It is actually a bit mask that reveals a number of things about the state of the kernel. These are documented in `linux/Documentation/sysctl/kernel.txt`.

Kernel Modules and the GPL

Kernel modules have a somewhat "special" relationship to the GPL. Linux is, as we all know, released under the GNU GPL, specifically version 2 of the GPL.

Everyone agrees that applications that run in User Space and only make use of published kernel APIs are not "derivative works" as defined by the GPL and therefore can remain proprietary. On the other hand, any code that is linked statically into the kernel is making use of GPL code and, by definition, is a derivative work and must be released under the GPL with source code available.

Dynamically loaded modules are a middle ground. Despite the fact that they are also making use of GPL code in the kernel, modules need not be open source, although as we've seen, the kernel will at least slap your wrist if you run a non-Open Source module. Nevertheless, there are kernel developers who would like to ban non-Open Source modules, and there was talk at one point of not allowing such modules to load.

In response, Linus asserted that banning proprietary modules would kill the commercial viability of Linux because hardware vendors have a legitimate interest in keeping their drivers proprietary. Releasing driver source code might reveal details of their proprietary hardware. If they were forced to release their source code, they simply wouldn't support Linux.

This has implications for how you treat your own device drivers. Technically speaking, there's often little reason to use modules in an embedded environment because the hardware is fixed and well-defined. So it would make sense to build the drivers into the kernel image. But if you want to maintain those drivers as proprietary, you're forced to build and release them as modules only.

Building Kernel Modules

The build process for kernel modules is somewhat more involved than the makefiles we've encountered up until now. The complication arises because modules must be built within the context of the kernel itself, specifically the kernel you intend to run the module with. The upshot of this is that, if you're building a module outside of the kernel tree as we are now, the makefile has to change into the top-level kernel source directory and invoke the makefile there. The actual build occurs in the top level of the kernel source tree with the results put back into the directory with our module source.

What happens is that the makefile in the `hello/` directory invokes make again changing directories to the root of the kernel source tree and passing in the location of `hello.c` as an argument. Take a look at the makefile for `hello`. Notice the first line:

```
obj-m := hello.o
```

Without going into a lot of detail, this tells the kernel build system to create a module from `hello.o`. Next we specify some debugging flags as necessary.

The really interesting stuff happens in line 20 where we invoke make on the `modules` target. The phrase:

```
-C /lib/modules/$(shell uname -r)/build
```

is how `make` changes to the kernel source tree. Execute the shell command `uname -r` and you'll get a numeric string representing the kernel version. Go to `/lib/modules` and you'll find a subdirectory of the same name. There is a link named `build`, which in fact links to the source tree for the currently executing kernel. You've no doubt guessed that `-C` means change directories. Finally, the phrase `M = $(shell pwd)` passes the current directory as the environment variable `M`.

The kernel source tree need not be the full kernel sources. Specifically, the `*.c` files are not required. What is required are the header, configuration, and make files. This is the normal format for popular desktop distributions. Take a look at the source tree for your kernel to see what's in there.

The end result is to compile `hello.c` into `hello.o` and then turn it into the loadable module `hello.ko`. You'll note that in the process, it creates a number of temporary files and a directory. Hence the `clean` target. If a module is built from two or more source files, the first line of the makefile expands to something like:

```
obj-m := module.o
module-objs := file1.o file2.o
```

When creating makefiles for any kind of project, whether it be a kernel module or a user space application, it's generally a good idea to start with an existing model. The makefile described here is a good starting point for other kernel module projects.

Finally, there's a point we glossed over when creating the Eclipse project for `hello`, the include paths. Kernel modules get their header files from the kernel source tree, specifically,

`<linux>/include` where `<linux>` is the absolute path to the kernel source. Use Project Properties to set the include path correctly. Refer back to Chapter 7 if you've forgotten how to do this.

The Role of a Module

As an extension of the kernel, a module's role is to provide some service or set of services to applications. Unlike an application program, a module does not execute on its own. Rather, it patiently waits until some application program invokes its service.

But how does that application program gain access to the module's services? That's the role of the `module_init()` function. By calling one or more kernel functions, `module_init()` "registers" the module's "capabilities" with the kernel. In effect, the module is saying "Here I am and here's what I can do."

Figure 12.1 illustrates graphically how this registration process works. The `init_module()` function calls some `register_capability()` function passing as an argument a pointer to a structure containing pointers to functions within the module. The `register_capability()` function puts the pointer argument into an internal "capabilities" data structure. The system then defines a protocol for how applications get access to the information in the capabilities data structure through system calls.

A specific example of registering a capability is the function `cdev_add()` that registers the capabilities of a "character" device driver. Subsequently, an application program gains access to the services of a character device driver through the `open()` system call. This is described in more detail shortly.

If a module is no longer needed and is removed with the `rmmod` command, the `module_exit()` function should "unregister" the module's capabilities, freeing up the entry in the capabilities data structure.

What's a Device Driver Anyway?

There are probably as many definitions of a *device driver* as there are programmers who write them. Fundamentally, a device driver is simply a way to "abstract out" the details of peripheral hardware so the application programmer doesn't have to worry about them.

In simple systems, a driver may be nothing more than a library of functions linked directly to the application. In general purpose OSs, device drivers are often independently loaded programs that communicate with applications through some OS-specific protocol. In multitasking systems like Linux, the driver should be capable of establishing multiple "channels" to the device originating from different application processes. In all cases,

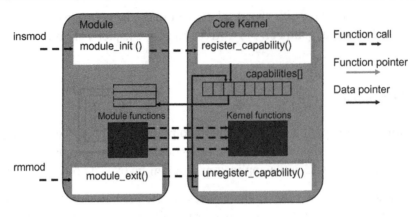

Figure 12.1
Linking a Module with the Kernel.[2]

though the driver is described in terms of an API that defines the services the driver is expected to support.

The device driver paradigm takes on additional significance in a protected mode environment such as Linux. There are two reasons for this. First, User Space application code is normally not allowed to execute I/O instructions. This can only be done in Kernel Space at Privilege Level 0. So a set of device driver functions linked directly to the application simply won't work. The driver must execute in Kernel Space. Actually, there are some hoops you can jump through to allow I/O access from User Space but it's better to avoid them.

The second problem is that User Space is swappable. This means that while an I/O operation is in process, the user's data buffer could get swapped out to disk. And when that page gets swapped back in, it will very likely be at a different physical address. So data to be transferred to or from a peripheral device must reside in Kernel Space, which is not swappable. The driver then has to take care of transferring that data between Kernel Space and User Space.

Linux Device Drivers

Chapter 9 included a brief discussion of device drivers. Let's review. Unix, and by extension Linux, divides the world of peripheral devices into four categories:

- Character
- Block
- Pipe
- Network

[2] Rubini, Alessandro, Jonathan Corbet, and Greg Kroah-Hartman, *Linux Device Drivers, third Ed.*, O'Reilly, 2005, p. 19.

The principal distinction between character and block is that the latter, such as disks, are randomly accessible, that is, you can move back and forth within a stream of characters. Furthermore, data on block devices are transferred in one or more blocks at a time and prefetched and cached. With character devices, the stream generally moves in one direction only. You can't for example go back and reread a character from a serial port. Some character devices may implement a seek function. Block devices must have a file system associated with them whereas character devices don't.

Pipes are pseudodevices that establish unidirectional links between processes. One process writes into one end of the pipe and another process reads from the other end.

Network devices are different in that they handle "packets" of data for multiple protocol clients rather than a "stream" of data for a single client. Furthermore, data arrives at a network device asynchronously from sources outside the system. These differences necessitate a different interface between the kernel and the device driver.

The /dev **Directory**

While other OSes treat devices as files, sort of, Linux goes one step further in actually creating a directory for devices. Typically, this is /dev. In a shell window on the workstation, cd /dev and ls −l s*. You'll get a number of interesting entries. Notice the first character in the flags field. "b" represents a "block" device and "c" represents a "character" device. Remember from Chapter 3 that "l" represents a link and "d" is a directory.

Between the Group field and the date stamp of the block and character device entries, where the file size normally appears, there are a pair of numbers separated by a comma. These are referred to as the "major device number" and the "minor device number" respectively. The major device number identifies the device driver. The minor device number is used only by the driver itself to distinguish among possibly different types of devices the driver can handle.

Prior to the 2.6 series, the major and minor device numbers were each eight bits, and there was a one-to-one correspondence between major numbers and device drivers. In effect, the major number was an index into a table of pointers to device drivers. Now a device number is a 32-bit integer, typedef'd as dev_t, where the high-order 12 bits are the major number and the remaining bits are the minor number. The registration mechanism now allows multiple drivers to attach to the same major number, provided they use different ranges of minor numbers.

Entries in /dev are created with the mknod command as, for example,

```
mknod /dev/ttyS1 c 4 65
```

This creates a /dev entry named ttyS1. It's a character device with major device number 4 and minor device number 65. There's no reference to a device driver here. All we've done with mknod is create an entry in the file system so application programs can "open" the device. But before that can happen we would have to "register" a character device driver with major number 4 and minor number 65 to create the connection.

The Low-Level User Space I/O APIs

We covered the low-level User Space APIs for accessing the features of device drivers fairly well in Chapter 9. Review if necessary.

Internal Driver Structure

The easiest way to get a feel for the structure of a device driver is to take a look at the code of one. There are a couple of simple examples in root_qtopia/home/src/simple_hw.

Create a new Eclipse makefile project in the simple_hw/ directory. You'll need to add the appropriate include path, which is in the Linux kernel source tree. See the Makefile for details. Finally, you'll need to create a new make target, modules.

Open simple_hw.c in the Eclipse editor. Basically, a driver consists of a set of functions that mirror the functions in the low-level User Space API. However, the driver's functions are called by the kernel, in Kernel Space, in response to an application program calling a low-level I/O function.

Driver Data Structures

Before we examine the functions in a device driver, we should look at some of the more important data structures that drivers make use of. With the exception of struct cdev, all of the structures described here are found in include/linux/fs.h.

struct file_operations: The file_operations structure is a collection of pointers to all of the possible functions that constitute the device driver APIs. This fulfills the requirement described in Figure 12.1 to communicate the capabilities of a module to the rest of the kernel. There are a lot of functions, but a driver need only declare the ones it actually implements. The unused entries are set to NULL and either cause some default action or an error if they are invoked.

The file_operations structure for simple_hw is at line 180. This uses the *tagged field* format for initializing the structure. Only the fields that we use are specified by name. All the others get set to zero or NULL. The owner field is almost always set to THIS_MODULE.

struct `file`: This structure represents an open file and is passed to most of the functions in the `file_operations` table. It is created when a User Space process calls `open()` and before the driver's `open()` function is called. About the only field of interest to driver writers is `void* private_data`. This allows you to allocate a data structure on a "per connection" basis and pass it around to all of the driver functions.

struct `inode`: This structure represents an entry in the file system. For hardware device drivers, it generally represents the node in `/dev` that is associated with the driver. It is passed to the `open()` and `release()` functions, among others, but my experience is that you rarely need access to anything in the `inode` structure.

struct `cdev`: This structure is defined in `include/linux/cdev.h` and is used to register the driver with the kernel. The fields of interest to driver writers are `ops`, which is initialized with the `file_operations` structure, and `owner` set to `THIS_MODULE`.

init() **and** *exit()*

Starting at line 36 is a set of four module parameters for major device number, transfer size—1, 2, or 4 bytes, offset within a 4-byte register and a base physical address. While this is a general purpose driver capable of reading and writing any of the GPIO ports, the default parameter values expose the LED bits.

Now let's go down to the `hw_init()` function at line 207. The first thing it does is sanity check, the offset parameter to make sure it's sensible for the selected transfer size. Then it gains exclusive access to the relevant I/O memory space by calling `request_mem_region()`, which returns a non-NULL value if it succeeds. The arguments to `request_mem_region()` are a base physical address, the number of ports to allocate, and the name of the device. `NR_PORTS` is passed in through the `Makefile` and defaults to eight. The device name is used to identify the driver in various `/proc` files.

Remember that the kernel deals in *virtual* addresses. So before we can access the GPIO ports, we have to map them into virtual address space. That's done with the `ioremap()` function.

Next we need to allocate a range of minor device numbers to represent our GPIO ports. These then belong to a major device number that serves as the link to our driver. There are two ways to do this. `register_chrdev_region()` takes a `dev_t` with a specific major/minor starting point, a number of minors, and a device name. It returns a negative error code if the specified range is not available. Alternatively, `alloc_chrdev_region()` will dynamically allocate a major number given a starting minor, the number of minors, and a device name.

The macro MKDEV(int, int) makes a dev_t out of a major and minor device numbers. There are also macros for retrieving the major and minor elements of a dev_t: MAJOR(dev_t) and MINOR(dev_t). Good practice dictates using these macros rather than manipulating the bit fields explicitly. There's no guarantee that the bit assignments for major and minor will remain the same in later kernel revisions. Besides, the code is more readable.

Having allocated a major device number, we can now register our driver with the kernel. This involves two steps:

1. Allocate or initialize a cdev data structure with cdev_alloc() or cdev_init().
2. Tell the kernel about it with cdev_add().

cdev_alloc() dynamically allocates and initializes a cdev structure. You must initialize the owner and ops fields before calling cdev_add(). Finally, we register the device driver with a call to cdev_add(). The arguments are:

- Pointer to the cdev structure.
- The based device number we got from alloc_chrdev_region().
- The number of devices we're registering, in this case eight.

In many cases, your device will have its own structure into which you may want to embed the cdev structure. Then you call cdev_init() passing in the address of the cdev structure and a pointer to the file_operations structure. You still need to initialize the owner field.

If hw_init() succeeds, the return value is zero. A non-zero return value indicates an error condition and the module is not loaded. Note that if hw_init() fails, any resources successfully allocated up to the point of failure must be returned before exiting. In this case, failure to register the device requires us to release the I/O region.

As might be expected, hw_cleanup() simply reverses the actions of hw_init(). It unregisters the device and releases the I/O region.

open() *and* release()

Move back to the top of the file at line 61. The kernel calls hw_open() when an application calls open() specifying /dev/<io_port> as the path. open() gets a struct inode and a struct file. open() usually does some sort of device and data structure initialization. In this case, there's nothing for it to do. We could have left it out and simply let the kernel take the default action, but it's useful to see the function prototype since open() is used in most drivers.

Interestingly enough, the driver equivalent of close is called release. Again, in this case there's nothing to do.

read() *and* write()

Moving down the file, we come to the interesting functions, hw_read() and hw_write(). The arguments to read and write are:

- A struct file.
- A pointer to a data buffer in *User Space.*
- A count representing the size of the buffer.
- A pointer to a file position argument. If the function is operating on a file, then any data transferred will change the file position. It is the function's responsibility to update this field appropriately.

A positive return value from either function is the number of bytes successfully transferred while a negative return is an error.

Both functions follow the same basic pattern. We extract the minor device number from a field of struct file and use that to set the register address. Then we adjust the count to be a multiple of the transfer size and allocate a buffer for the data. Note that kmalloc() is the kernel equivalent of malloc() in User Space. At this point, hw_write() calls copy_from_user() to transfer the User Space data to the local buffer. We then drop into a loop to carry out transfers of the specified size. Finally, hw_read() calls copy_to_user() to transfer the local buffer to User Space. copy_from_user() and copy_to_user() encapsulate the details of moving data between Kernel Space and User Space.

You may be wondering about the ioreadx() and iowritex() calls. Can't we just treat the virtual address as an ordinary C variable? In most cases we can, but there are some architectures where, for whatever reason, this doesn't work. So in the interest of portability, you should use the functions. hw_write() does show how to access the addresses directly.

hw_write() has a couple of optimizations for the LEDs. The data is inverted so that a 1 signifies a lit LED. The data is shifted five bits left so the least significant four bits represent the LEDs.

Figure 12.2 attempts to capture graphically the essence of driver processing for a write operation. The user process calls write(). This invokes the kernel through an INT instruction where the write operation is conveyed as an index into the syscall[] table. The filedes argument contains, among other things, a major device number so that the kernel's write function can access the driver's file_operations structure. The kernel calls the driver's write function, which copies the User Space buffer into the Kernel Space.

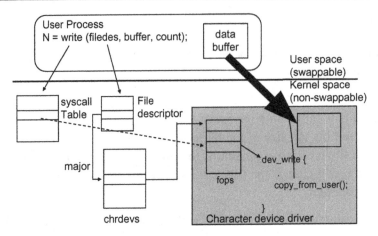

Figure 12.2
Processing a Write Call.

Building and Running the Driver

Execute `make` in the `simple_hw/` directory to build the module. We're using the dynamically allocated major number mechanism, so the question now is, how do we know what major number was allocated so we can create the corresponding nodes in `/dev`? Have a look at the `hw_load` script. It begins by installing the module. Then it lists the file `/proc/devices`, which shows the connection between major device numbers and drivers. That output is then piped to the pattern scanning program `awk` that looks for our module name and returns the associated major number.

Then the script creates eight device nodes whose names match the register names in the S3C2440 data sheet.

Do the following on the target board:

1. `cd /home/src/simple_hw`.
2. `cat /proc/iomem`   Verify that the range 0x56000010 to 0x5600002f is *not* listed. This means it's available for our driver.
3. `./hw_load`.
4. `cat /proc/iomem`   0x56000010 to 0x5600002f is attached to `simple_hw`.
5. `cat /proc/devices`   Shows which major number was allocated to `simple_hw`.
6. `ls −l /dev/pio*`   Shows the device nodes created by `hw_load`.
7. `lsmod`   Shows that `simple_hw` is loaded.
8. `echo −n "1" > /dev/pio_GPBDAT`   One LED should be lit.

`echo` normally appends an end-of-line character (0x10) to every string it outputs. But the only thing you'll see on the LEDs is the last character in the string. The "-n" option does not append the end-of-line, so you see the actual last character in the string. Try few more patterns. When you're finished, execute `./hw_unload` to remove the module and the device nodes. Of course, you could always just reboot.

"Big deal," you may say. "We blinked LEDs back in Chapter 7." True, but what this exercise shows is that by making our device accessible through the standard driver mechanism, we can access it through standard Linux utilities. We can access it from scripts. In fact, I challenge you to write a script that duplicates the behavior of the led program from Chapter 7.

Debugging Kernel Code

Kernel Space code presents some problems in terms of debugging. To begin with, Eclipse and GDB rely on kernel services. If we stop the kernel, at a breakpoint for example, those services aren't available. Consequently, some other approaches tend to be more useful for kernel and module code.

printk

Personally, I've never been a big fan of the `printf()` debugging strategy. On occasion, it's proven useful but more often than not the print statement just isn't in the right place. So you have to move it, rebuild the code, and try again. I find it much easier to probe around the code and data with a high-quality source-level debugger.

Nevertheless, it seems that many Unix/Linux programmers rely heavily on `printf()` and its kernel equivalent `printk()`. At the kernel level this makes good sense. Keep in mind of course that `printk()` exacts a performance hit.

While `printk()` statements are useful during development, they should probably be taken out before shipping production code. Of course, as soon as you take them out, someone will discover a bug or you'll add a new feature and you'll need them back again. An easy way to manage this problem is to encapsulate the `printk()`'s in a macro as illustrated in Listing 12.1.

While you're debugging, define the `DEBUG` macro on the compiler command line. When you build production code, `DEBUG` is not defined and the `printk()` statements are compiled out.

/proc *Files*

We first encountered /proc files back in Chapter 3. The `/proc` filesystem serves as a window into the kernel and its device drivers. `/proc` files can provide useful runtime information and can also be a valuable debugging tool. In fact, many Linux utilities, `lsmod` for example, get

```
#ifdef DEBUG
#define PDEBUG(fmt, args…) printk (<1> fmt, ## args)
#else
#define PDEBUG(fmt, args…)                    // nothing
#endif
```

Listing 12.1
Encapsulate `printk()` in a Macro.

```
#include <linux/proc_fs.h>

struct proc_dir_entry *create_proc_read_entry (char *name,
                         mode_t mode, struct proc_dir_entry *base,
                         read_proc_t *read_proc, void *data);

int read_proc (char *page, char **start, off_t offset,
                         int count, int *eof, void *data);
```

Listing 12.2
`/proc` file functions.

their information from `/proc` files. Some device drivers export internal information via `/proc` files and so can yours.

`/proc` files are said to be harder to implement than `printk()` statements but the advantage is that you only get the information when you ask for it. Once a `/proc` file is created, there's no runtime overhead until you actually ask for the data. `printk()` statements on the other hand always execute.

A simple read-only `/proc` file can be implemented with the two function calls shown in Listing 12.2. A module that uses `/proc` must include `<linux/proc_fs.h>`. The function `create_proc_read_entry()` creates a new entry in the `/proc` directory. This would typically be called from the module initialization function. The arguments are:

- Name. The name of the new file.
- File permissions, `mode_t`. Who's allowed to read it. The value 0 is treated as a special case that allows everyone to read the file.
- `struct proc_dir_entry`. Where the file shows up in the `/proc` hierarchy. A NULL value puts the file in `/proc`.
- Read function. The function that will be called to read the file.
- Private data. An argument that will be passed to the read function.

The `read_proc()` function is called as a result of some process invoking `read()` on the `/proc` file. Its arguments are:

- `page`. Pointer to a page (4 KB) of memory allocated by the kernel. `read_proc()` writes its data here.
- `start`. Where in page `read_proc()` starts writing data. If `read_proc()` returns less than a page of data you can ignore this and just start writing at the beginning.
- `offset`. This many bytes of page were written by a previous call to `read_proc()`. If `read_proc()` returns less than a page of data you can ignore this.
- `count`. The size of `page`.
- `eof`. Set this non-zero when you're finished.
- `data`. The private data element passed to `create_proc_read_entry()`.

Like any good read function, `read_proc()` returns the number of bytes read.

As a trivial example, we might want to see how many times the read and write functions in `simple_hw` are called. Listing 12.3 shows how this might be done with a `/proc` file.

Handling Interrupts

Often, I/O operations take a long time to complete or else the computer must wait for some event to occur. In these cases, we would like a way to signal the computer that the operation is complete or the event has occurred. This is known as the *interrupt*. Interrupts can be thought of as a hardware equivalent of signals in User Space. They are asynchronous events for which the driver registers a handler. When the interrupt occurs, the handler is invoked.

```
Add to simple_hw.c

#include <linux/proc_fs.h>

int read_count, write_count;       // These are incremented each time the read and
                                    //         write functions are called.

int hw_read_proc (char *page, char **start, off_t offset, int count, int *eof, void *data)
{
            return sprintf (page, "simple_hw.  Read calls: %d, Write calls: %d\n",
                        read_count, write_count);
}

In hw_init ()

            create_proc_read_entry ("simple_hw", 0, NULL, hw_read_proc, NULL);
```

Listing 12.3
Adding a `/proc` file to a module.

Interrupt handlers, by their nature, run concurrently with other code. Thus, they inevitably raise issues of concurrency and contention for data structures and hardware. All of the tools and techniques for handling race conditions and concurrent access come into play when dealing with interrupt handlers.

Although the driver code we've encountered up to now runs in the context of whatever process called it, interrupt handlers, being asynchronous, run outside the context of any process in a state known as *interrupt context*. As a result, they are rather limited in what they can do. Here are some of the things that interrupt handlers are not allowed to do:

- *Access user space*. Since the interrupt handler is executing asynchronously with respect to processes, we don't know what process is running when the handler is invoked.
- *Sleep*. Only process context can sleep.
- *Execute for a long time*. The handler *interrupted* something. We want to get back to whatever that was as soon as possible. If an interrupt handler does require lengthy operation, it should invoke a *bottom half*.

The example for experimenting with interrupts is called `simple_int.c`. The operation of `simple_int` is quite straightforward. We connect a GPIO output pin to a GPIO input pin that's also configured to generate interrupts on a rising edge. Figure 12.3 shows how to use a jumper plug to connect the appropriate bits together.

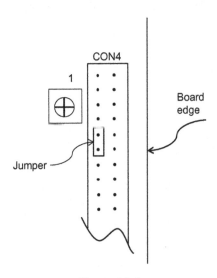

Figure 12.3
Connecting Output Bit to Input Bit.

The driver's write function toggles the output bit for each byte sent to it. Thus, every second byte causes an interrupt. The interrupt handler gets the current time with microsecond resolution and writes a timestamp string to a circular buffer. The read function "sleeps" until the interrupt handler wakes it up. Then it passes the contents of the circular buffer back to User Space. The result is a string of timestamps that show how quickly the system can generate interrupts. This operation is illustrated graphically in Figure 12.4.

Registering an Interrupt Handler

Like any driver resource, an interrupt handler must be registered with the kernel before it can be used. Open simple_int.c with an editor. Starting at line 436 in the module init function, there's a set of four calls to request_irq() representing different operating modes. We'll come back to the tasklet and wq modes later. For now we'll register my_interrupt as the interrupt handler. The arguments to request_irq() are:

- The interrupt number you're registering.
- A pointer to the handler function, which takes two arguments to be described shortly.
- A flags bit mask.
- The device name. This is used by /proc/interrupts to identify the devices holding interrupts.
- A pointer used for shared interrupt lines. Typically, this would point to the driver's private data area as a way of identifying which device caused the interrupt. If the interrupt isn't shared, this can be NULL.

As usual, the return value is 0 if the function succeeds and a negative error code if it fails. The most likely error is –EBUSY, meaning that another device has registered this non-shared interrupt.

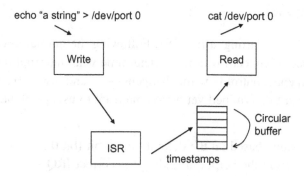

Figure 12.4
Operation of simple_int.

`flags` has several bits including these three:

- `IRQF_DISABLED`—indicates a "fast" handler when set. A fast handler executes with interrupts disabled. This bit should not be set unless it's absolutely necessary.
- `IRQF_SHARED`—when set, it indicates that the interrupt line is shared. Multiple registrations of a shared interrupt succeed.
- `IRQF_SAMPLE_RANDOM`—indicates that the generated interrupts can contribute to the "entropy pool" used by `/dev/random` and `/dev/urandom`.

The last flag deserves a more detailed explanation. `/dev/random` and `/dev/urandom` generate truly random numbers used to create secure keys for encryption. These random numbers are based on random events that contribute to an "entropy pool." If your device generates interrupts at truly random intervals, a keyboard for example, you should set this bit.

However, devices that generate interrupts at regular intervals, the time tick or the vertical blanking signal, shouldn't set this bit. Interestingly, network drivers shouldn't set this bit because they are subject to hacking in the form of regular and predictable packet timing from determined hackers. The file `drivers/char/random.c` has some relevant comments on this issue.

When the kernel calls the interrupt handler in response to an interrupt, it passes two arguments:

- The interrupt number that was activated.
- The private data argument passed to `request_irq()`.

The handler's return value indicates whether or not it actually serviced the interrupt. If the handler services the interrupt, it should return `IRQ_HANDLED`. Otherwise, it should return `IRQ_NONE`. This helps the kernel deal with shared interrupts and detect spurious interrupts.

Probing for the Interrupt

Take a look at `int_init()` starting at line 381. Following the normal device registration process, the next issue of concern is how we determine what interrupt number to request. Two load time arguments control how this happens—`irq` and `probe`. `irq` is initialized to −1, meaning that we haven't determined yet which interrupt to use. `probe` has three possible values:

- 0—use a default value based on the default output bit (bit 0).
- 1—get some help from the kernel in finding the correct IRQ.
- 2—the driver will do the probing itself.

The kernel provides a pair of functions that can help determine which interrupt line a device is connected to. These are illustrated in the function `kernel_probe()` beginning at line 276.

`probe_irq_on()` returns a bit mask of currently unassigned interrupts. The function keeps the return value and then arranges for the device to generate an interrupt. It then calls `probe_irq_off()`, passing the mask value that was returned by `probe_irq_on()`. `probe_irq_off()` returns a number representing the interrupt that was generated after the call to `probe_irq_on()`. If no interrupt occurred, it returns 0. If multiple interrupts occurred, an ambiguous condition, it returns a negative number.

The final option is known as do it yourself probing. This is illustrated in `self_probe()` starting at line 322. It is really just a variation on the kernel-assisted method described above. In this case, we know what interrupts are possible based on possible combinations of output bits and input interrupts that can be jumpered, so we attach a probing interrupt handler to any of them that are free. Then we arrange to generate an interrupt and see what happens.

Try it Out

Install a jumper plug as described in Figure 12.4. `simple_int.ko` should already exist having been built by the same make command that built `simple_hw.ko`. Execute the following commands as root user in the `simple_hw/` directory:

- If `simple_hw` is currently loaded, execute `./hw_unload`.
- `./int_load`.
- `echo "some string of text" > /dev/port0`.
- `cat /dev/port0`.

You should see a series of timestamp messages, about 10 μs apart, one for each two bytes in the input string.

- `rmmod simple_int`.
- `insmod simple_int.ko probe = 1`.

After the above two commands, `simple_int` will have the same major device number. Repeat these two commands with `probe = 2`.

Here's a programming challenge. Modify `simple_int.c` so that rather than reporting the interrupt interval, it reports interrupt latency, the time from setting the interrupt line to the beginning of the interrupt handler.

Deferred Processing—the "Bottom Half"

An interrupt service routine (ISR) should execute quickly so that its device's interrupt can be re-enabled. Also, we would like to get back into a process context as soon as possible so we have a better handle on the system's state.

Virtually, all useful OSs split interrupt handling into two parts. The first part is the ISR itself that responds directly to the interrupt. At the very least, the ISR needs to clear the device interrupt flag and probably read or write some data. But that's all it should do. If there's more extensive work to be done, processing a network packet for example, that should be deferred to some time later when it's more "convenient" for the system.

In Linux, these two parts are called the "top half" and the "bottom half." The bottom half mechanism has undergone extensive revision and upgrade over the years and, in fact, the original bottom half mechanism is deprecated in the 2.6 series. The two strategies currently available for bottom half processing are *tasklets* and *workqueues*.

A tasklet is simply a function taking a long data argument that is scheduled to run at some time in the future. It executes in "soft interrupt" context meaning that interrupts are enabled but it must still observe the rules of ISRs.

Workqueues are very much like tasklets except that they operate in the context of a dedicated "kernel thread." This means, among other things, that the workqueue function can sleep. The function need not execute atomically.

`simple_int` can implement either top half/bottom half paradigm using a pair of module parameters. Alternate interrupt handlers are registered when either `wq` or `tasklet` are set non-zero.

The "top half" interrupt handler writes a binary timestamp, two long words, to a circular buffer and invokes the bottom half, which happens to be the same function for both the tasklet and workqueue versions. Writing a binary timestamp is clearly faster than formatting and writing ASCII text.

The handler also increments a counter so we know how many times it fired off before the bottom half got to execute.

The bottom half function, `do_bottom_half()` at line 200, first reads and resets the interrupt count. It formats the count as text and writes it to a text buffer. Next, it reads whatever is in the binary timestamp buffer, formats that as text, and writes it to the text buffer. Finally, it wakes up the read function. Figure 12.5 depicts this graphically.

Try it Out

- `rmmod simple_int`
- `insmod simple_int.ko tasklet = 1`
- `echo "a string of text" > /dev/port0`
- `cat /dev/port0`
- `rmmod simple_int`

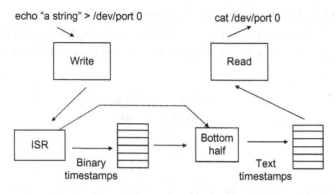

Figure 12.5
`simple_int` with Bottom Half.

- `insmod simple_int.ko wq = 1`
- `echo "a very long string of text" > /dev/port0`
- `cat /dev/port0`

What do you notice? The tasklet tends to execute fairly quickly. Workqueues being under the control of the scheduler take rather longer.

Building Your Driver into the Kernel

In a production-embedded environment, there is little reason at all to use loadable kernel modules unless there are proprietary considerations. So when you have your new driver working, you'll probably want to build it into the kernel executable image. This means integrating your driver into the kernel's source tree. It's not as hard as you might think.

To begin with, you need to copy your driver source files into the kernel source tree. Let's do that with `simple_hw.c`. In the absence of any definitive documentation or recommendations on this issue, I recommend putting the driver source in `usr/src/linux/drivers/char`.

The source file requires a couple of minor modifications. In the declaration of the module initialization function, add "__init" just before `hw_init` so that it looks like this:

```
int __init hw_init (void)
```

`__init` causes the function to be compiled into a special segment of initialization functions that are called as part of the boot up process. Once boot up is complete, this segment can be discarded and the memory reused.

When the module is built into the kernel, the function identified by `module_init()` is added to a table of initialization functions that are called at boot up. The upshot is that you don't have to modify `main.c` to add a new initialization function every time you add a new driver. It happens automatically.

Also, we no longer need `hw_cleanup()` because the device will never be removed. You can bracket that function and the `exit_module()` macro with:

```
#ifdef MODULE
#endif
```

Next you'll need to edit the Makefile in `usr/src/linux/drivers/char`. Note that the Makefile consists mostly of lines of the form:

```
obj-$(CONFIG_yyyy) + = yyyy.o
```

where `CONFIG_yyyy` represents an environment variable set by the `make xconfig` process and `yyyy.o` is an object file. Remember from the last chapter that all of these environment variables end up with one of two values:

"y" Yes. Build this feature into the kernel.
"m" Module. Build this feature as a kernel-loadable module
or it's commented out.

When the top-level Makefile completes its traverse of the kernel source tree, the environment variable `obj-y` contains a list of all the object files that should be linked into the kernel image, and `obj-m` is a list of the object files that should be built as kernel-loadable modules.

Add the following line to the Makefile, perhaps at the end:

```
obj-$(CONFIG_SIMPLE) + = simple_hw.o
```

You also need to add an entry into the configuration menus. Open the `Kconfig` file in `/usr/src/arm/linux/drivers/char`. Using the entries for the mini2440 device drivers as a model, create an entry for `CONFIG_SIMPLE`.

Go back up to `/usr/src/arm/linux` and run `make xconfig`. Scroll down to the character device section and note that your new entry shows up and has (NEW) after the prompt. `xconfig` has noticed that there's a configuration variable in the menu that's not yet in the `.config` file.

This concludes our whirlwind tour of programming in kernel space. In the next chapter, we'll look at BusyBox and system initialization.

Resources

Rubini, Alessandro, Jonathan Corbet, and Greg Kroah-Hartman, *Linux Device Drivers, third Ed.*, O'Reilly, 2005. On the one hand, this is a very readable and thorough treatment of kernel-level programming. On the other hand, it's getting rather dated as it describes kernel version 2.6.10. Consequently, many of the details are off.

The subject of module and device driver programming is way more extensive than we've been able to cover here. Hopefully, this introduction has piqued your interest and you'll want to pursue the topic further:

Venkateswaran, Sreekrishnan, *Essential Linux Device Drivers*, Prentice-Hall, 2008. This one deals with kernel version 2.6.23/24.

## PART 3

Components and Tools

## CHAPTER 13

BusyBox and Linux Initialization

Linux: the choice of a GNU generation
ksh@cis.ufl.edu put this on T-shirts in 1993

Very often the biggest problem in an embedded environment is the lack of resources, specifically memory and storage space. As you've no doubt observed, either in the course of reading this book, or from other experience, Linux is *big*! The kernel itself is often in the range of 2–3 MB, and then there's the root file system with its utility programs and configuration files. In this chapter, we'll look at a powerful tool for substantially reducing the overall "footprint" of Linux to make it fit in limited resource embedded devices.

The other topic we'll address in this chapter is User Space initialization and specifically the question of how to get our thermostat application to start at boot up. We'll also add a simple display to the thermostat.

Introducing BusyBox

Even if your embedded device is "headless," that is it has no screen and/or keyboard in the usual sense for user interaction, you still need a minimal set of command-line utilities. You'll no doubt need `mount`, `ifconfig`, and probably several others to get the system up and running. Remember that every shell command-line utility is a separate program with its own executable file.

The idea behind BusyBox is brilliantly simple. Rather than have every utility be a separate program with its attendant overhead, why not simply write one program that implements *all* the utilities? Well, perhaps not all, but a very large number of the most common utilities. Most utilities require the same set of "helper" functionality such as parsing command lines and converting ASCII to binary. Rather than duplicating these functions in hundreds of files, BusyBox implements them exactly once.

The BusyBox project began in 1996 with the goal of putting a complete Linux system on a single floppy disk that could serve as a *rescue disk* or an installer for the Debian Linux distribution. A rescue disk is used to repair a Linux system that has become unbootable. This means the rescue disk must be able to boot the system and mount the hard disk file systems, and it must provide sufficient command-line tools to bring the hard disk root file system back to a bootable state.

DOI: http://dx.doi.org/10.1016/B978-0-12-415996-9.00013-7

Subsequently, embedded developers figured out that this was an obvious way to reduce the Linux footprint in resource-constrained embedded environments. So the project grew well beyond its Debian roots and today BusyBox is a part of almost every commercial embedded Linux offering, and it is found in a wide array of Linux-based products including numerous routers and media players.

BusyBox calls itself the "Swiss army knife" of embedded Linux because, like the knife, it's an all-purpose tool. Technically, the developers refer to it as a "multicall binary," meaning that the program is invoked in multiple ways to execute different commands. This is done with symbolic links named for the various utilities. These links then all point to the BusyBox executable.

Configuring and Installing BusyBox

There is a version of BusyBox on the board's DVD. Although it's a little old, I suggest that it's a good place to start just because that's what's running on the board now. Later on you can get a more recent version from the web site. Copy `linux/busybox-1.13.3-mini2440.tgz` from the DVD to your home directory and untar it. You'll find a new subdirectory, `busybox-1.13.3/`.

BusyBox is highly modular and configurable. While it is capable of implementing over 300 shell commands, by no means are you required to have all of them in your system. The configuration process lets you choose exactly which commands will be included in your system. Table 13.1 is the full list of commands available in recent releases of BusyBox.

BusyBox supports the `xconfig` and `menuconfig` make targets for configuration just like the kernel. Figure 13.1 shows the `xconfig` menu while Figure 13.2 shows the top-level `menuconfig` menu. Also like the kernel there's a `make help` target that lists the other make targets. Oddly, `xconfig` doesn't show in the list.

There are no `*_defconfig` targets for creating default configurations, so our initial configuration will have to come from somewhere else. That's what the file `fa.config` is for. Copy it to `.config` and then execute either `make xconfig` or `make menuconfig`.

BusyBox Settings

While the xconfig navigation panel shows a number of option categories, there are really only two options: (1) to configure, build, and install BusyBox and (2) to select the desired applets. Under Build Options, you have the choice of building BusyBox either with statically linked libraries or shared libraries. A statically linked BusyBox is a much bigger executable, but then you don't need the shared libraries. As an example, my BusyBox with shared libraries is 634,368 bytes. The statically linked version is 1,701,544 bytes. But the /lib

Table 13.1: BusyBox Commands

[	dirname	hwclock	md5sum	resize	tac
[[	dmesg	id	mdev	rm	tail
acpid	dnsd	ifconfig	mesg	rmdir	tar
addgroup	dnsdomainname	ifdown	microcom	rmmod	taskset
adduser	dos2unix	ifenslave	mkdir	route	tcpsvd
adjtimex	dpkg	ifplugd	mkdosfs	rpm	tee
ar	du	ifup	mkfifo	rpm2cpio	telnet
arp	dumpkmap	inetd	mkfs.minix	rtcwake	telnetd
arping	dumpleases	init	mkfs.vfat	run-parts	test
ash	echo	inotifyd	mknod	runlevel	tftp
awk	ed	insmod	mkpasswd	runsv	tftpd
basename	egrep	install	mkswap	runsvdir	time
beep	eject	ionice	mktemp	rx	timeout
blkid	env	ip	modprobe	script	top
brctl	envdir	ipaddr	more	scriptreplay	touch
bunzip2	envuidgid	ipcalc	mount	sed	tr
bzcat	expand	ipcrm	mountpoint	sendmail	traceroute
bzip2	expr	ipcs	mt	seq	TRUE
cal	fakeidentd	iplink	mv	setarch	tty
cat	FALSE	iproute	nameif	setconsole	ttysize
catv	fbsplash	iprule	nc	setfont	udhcpc
chat	fdflush	iptunnel	netstat	setkeycodes	udhcpd
chattr	fdformat	kbd_mode	nice	setlogcons	udpsvd
chgrp	fdisk	kill	nmeter	setsid	umount
chmod	fgrep	killall	nohup	setuidgid	uname
chown	find	killall5	nslookup	sh	uncompress
chpasswd	findfs	klogd	od	sha1sum	unexpand
chpst	flash_lock	last	openvt	sha256sum	uniq
chroot	flash_unlock	length	passwd	sha512sum	unix2dos
chrt	fold	less	patch	showkey	unlzma
chvt	free	linux32	pgrep	slattach	unlzop
cksum	freeramdisk	linux64	pidof	sleep	unzip
clear	fsck	linuxrc	ping	softlimit	uptime
cmp	fsck.minix	ln	ping6	sort	usleep
comm	fsync	loadfont	pipe_progress	split	uudecode
cp	ftpd	loadkmap	pivot_root	start-stop-daemon	uuencode
cpio	ftpget	logger	pkill	stat	vconfig
crond	ftpput	login	popmaildir	strings	vi
crontab	fuser	logname	printenv	sty	vlock
cryptpw	getopt	logread	printf	su	volname
cut	getty	losetup	ps	sulogin	watch
date	grep	lpd	pscan	sum	watchdog
dc	gunzip	lpq	pwd	sv	wc
dd	gzip	lpr	raidautorun	svlogd	wget
deallocvt	hd	ls	rdate	swapoff	which
delgroup	hdparm	lsattr	rdev	swapon	who
deluser	head	lsmod	readlink	switch_root	whoami
depmod	hexdump	lzmacat	readprofile	sync	xargs
devmem	hostid	lzop	realpath	sysctl	yes
df	hostname	lzopcat	reformime	syslogd	zcat
dhcprelay	httpd	makemime	renice		zcip
diff	hush	man	reset		

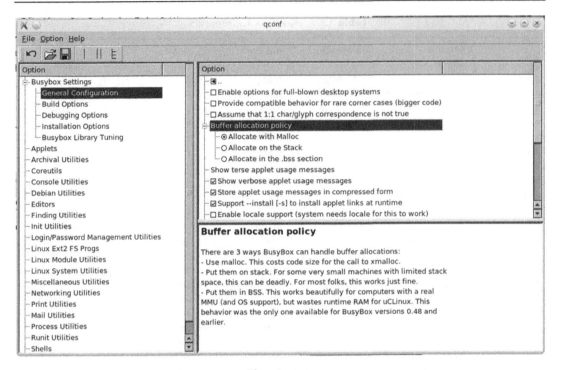

Figure 13.1
BusyBox xconfig Menu.

directory, where the shared libraries are kept, adds up to 14 MB. So the question is, do you need those shared libraries for anything else?

Build Options is also where you can select the cross-compiler prefix and note that it's already set for you.

Under Installation Options you have a choice of how the symbolic links are created. They can be soft links, hard links, or script "wrappers." The accompanying box describes the distinction between hard and soft links. Since hard links to the same data all share the same inode, this can be a useful option on systems with a limited number of inodes. Soft links are the default and usually the best way to go.

The Installation Prefix is wrong. This identifies the root of the file system into which you are installing BusyBox. In our case this is /home/target_fs. You can specify it here or on the command line for the make install step by adding CONFIG_PREFIX = /home/target_fs. Interestingly, in earlier releases of BusyBox, if you didn't specify a location, it would default to a directory within the BusyBox tree. That doesn't work anymore.

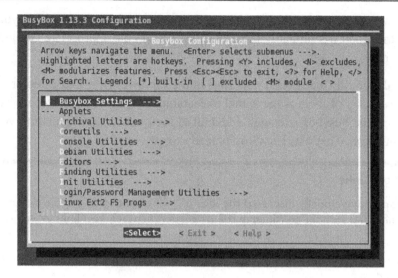

Figure 13.2
BusyBox menuconfig Top Menu.

Applets

The 300-odd BusyBox applets are divided among several categories. These categories correspond closely to the BusyBox directory tree. In many cases, the BusyBox version of a utility omits some of the more obscure options. You might think of these as "lite" versions. Nevertheless, the options that are available will suffice for just about anything you need in your embedded system.

In some cases, you can select the options you want for a command. For example, select the Coreutils category and scroll down to ls. You'll find several selectable ls options. Presumably, the reason for doing this is to reduce the code size if that's important. It would be nice if help gave some indication of how much space could be saved with each option.

Browse through the categories to get a feel of what's there. You'll find that there's an awful lot of stuff enabled that you probably don't need, like the vi editor, for example. And do you really need tar and gzip on an embedded device? Take a look at the Shells category. The default shell is ash, claimed to be the most complete and "pedantically correct" shell in the BusyBox distribution. Two other shells, hush and msh, offer less functionality in exchange for a smaller footprint.

Building and Installing

BusyBox uses the normal pattern of make and make install commands to build and install the package. But if you execute make, you'll find that it fails with the following error message:

```
Makefile:1273: *** mixed implicit and normal rules. Stop
```

This is a known problem resulting from an update to make syntax. I could tell you how to fix it, but at this point it's probably more useful if you do the appropriate Google searches and fix it yourself.

Following make you'll find several new files in the top-level BusyBox directory all starting with "busybox." One of them is the actual executable, busybox. Next, execute make install, which will copy the busybox executable and all of the links to the destination you specified during configuration. That's it, BusyBox is ready to go.

Hard vs. Soft Links

Back in Chapter 3, we briefly discussed the idea of "links," entries in the file system that just point to other files, without considering the distinction between *hard* links vs. *soft* or *symbolic* links.

Try this. In a shell window on your workstation go to ~/target_fs/home/src/led. Execute these commands:

```
ln led.c led2.c
ln -s led.c led3.c
ls -l
stat led.c
stat led2.c
stat led3.c
```

What do you notice? In particular, note the reference counts, the number just after the permissions field, and the file sizes. led3.c is identified as a link that points to led.c. Its length is 5, the size of the file *name* it is pointing to. By contrast, led2.c and led.c have the same length and a reference count of 2 whereas everything else has a reference count of 1. The stat commands tell us what inode represents each of the files. The thing to notice is that led.c and led2.c share the same inode, but led3.c has a different inode.

Now do this:

```
rm led.c
ls -l
```

On my Fedora 14, the entry for led3.c is highlighted to show that the file it points to doesn't exist. The reference count for led2.c is now 1.

Finally, rename led2.c as led.c:

```
mv led2.c led.c
ls -l
```

The led3.c entry is once again pointing at a valid entry.

The point of this exercise is that hard links are alternate names, or aliases if you will, for the same *data*. They all point to the same data blocks on the disk. That data can't be deleted until the reference count goes to zero. By contrast, a soft link or symbolic link points, not at data, but at a file *name*, which may or may not exist.

Using BusyBox

The BusyBox executable is usually installed in the /bin directory where most user-level command-line utilities reside. Then, for example, in /bin we create the following symbolic link:

```
ln-s busybox ls
```

Now when you invoke ls from a shell, what really gets executed is BusyBox. The program figures out what it's supposed to do by looking at the first argument to main(), which is the name that invoked it, i.e., ls. A similar link is created for every command that BusyBox implements.

You can also invoke BusyBox directly, passing a command name and its arguments as a command line, like this:

```
busybox ls-l
```

Executing busybox —help or just busybox by itself yields a help message including a list of the commands built into the current instance of BusyBox.

A Thermostat Display

Our thermostat is now a relatively useful network-enabled device, but it has no local display. If your Mini2440 has an LCD, we can now remedy that problem. First a little history.

Many years ago, the typical "console" device for a Unix, and later Linux, system was a dumb terminal connected to the computer via a serial line. This is a classic case of a character device. Then video terminals like Digital Equipment Corporation's (DEC's) VT100 developed some smarts, providing means to move the cursor around, change colors, and selectively erase parts of the screen. Later, the PC introduced the notion of bit-mapped "framebuffer" devices built into the computer itself.

These developments inevitably introduced additional layers of complexity in the Linux driver model. During the era of serial line video terminals, every manufacturer initially invented its own protocol for things like cursor control. So for the sake of portability, these various, usually incompatible, protocols had to be "abstracted out" into something uniform.

The role of a "console device driver" then is to do exactly that—provide a uniform interface to applications that need to do more than just output a stream of characters. The VT100 protocol has emerged as the de facto standard for terminal control and this is what a typical console driver exposes as its API.

Likewise, framebuffer devices come in various shapes and sizes with a range of color depths and various ways of mapping bits and bytes to pixels on the screen. The role of a framebuffer driver is to expose a uniform API that allows applications (and indeed other drivers like the console) to manage these differences in a consistent manner.

The driver for the LCD panel is a framebuffer driver. Framebuffers are character devices with major number 29. Unfortunately, the framebuffer API is not very well documented. But that's OK because we're not going to use it directly anyway.

Instead, we're going to use a console driver that's attached to device number 4 0. This driver properly attaches itself to the S3C2410 LCD framebuffer driver such that anything written to it ends up displayed on the screen.

ANSI Terminal Escape Sequences

There is in fact a standard protocol for controlling a console device. It is based on the protocol originally developed by DEC for its VT100 series video terminals. The standard goes by a number of common names, most of which contain the acronym ANSI (American National Standards Institute).

Terminal control commands all begin with the escape character, 27 or $0 \times 1b$, usually abbreviated <ESC>. This is usually followed by a "[" character and one or more characters that define some control action. The string <ESC>[is known as the Control Sequence Introducer or CSI.

The complete ANSI standard for terminal control is quite extensive, and I have yet to find a genuinely readable version of it. For reference, the version written by the European Computer Manufacturers Association (ECMA) is included on the CD as ECMA-048.pdf. To find other versions, Google something like "ansi terminal escape."

Fortunately, there's not a whole lot we need to do with escape sequences for our purposes. The main thing we want to do is to move the cursor around to create a user-friendly display for our thermostat. The command to set the cursor is:

```
<ESC>[<row>;<col>H
```

where <row> and <col> are numbers representing the desired cursor position. Row 1, column 1 is the upper left-hand corner of the screen. So, for example, to set the cursor at row 2, column 12, we could create the following string in C:

```
"\x1b[2;12H"
```

Thermostat Display

With that background, let's add a display to our thermostat that looks something like Figure 13.3. Create a new Eclipse project called "lcd" located in `home/src/lcd`. Open `lcdutils.c`. The file currently has the following functions:

- `lcd_erase (int what)`—erases either the current line or the entire screen depending on the value of `what`. A pair of constants is defined in lcdutils.h. Other erase options could be added to this function.
- `int lcd_init (void)`—initialize the LCD screen. Opens a file to the console device and returns the file descriptor. Gets access to I/O port C to manipulate the backlight. If anything fails it returns −1. Note that the file descriptor is not normally needed but is available if necessary.
- `lcd_set_cursor (int row, int col)`—moves the cursor to the indicated position.
- `lcd_write_string (int row, int col, char *string)`—writes a string at the indicated position.
- `lcd_write_number (int row, int col, int value, int width)`—writes a decimal `value` at the indicated position with the indicated `width`.
- `lcd_backlight (int on)`—if `on` is zero, turn off the backlight. Otherwise turn it on.
- `lcd_close (void)`—closes the file to the console and unmaps the I/O space.

There's also a simple program there, `lcdtest.c`, that exercises these functions. Take a look at `lcdtest.c`. Then build it and try it out.

That's sufficient functionality to create the screen shown in Figure 13.3. Have a go at it using the net version of the thermostat. Then, when any of the parameters is changed from the network, you'll see it updated on the display. Create an Eclipse make target for `netthermo_t`.

And of course feel free to add new functionality to `lcdutils` as the need arises.

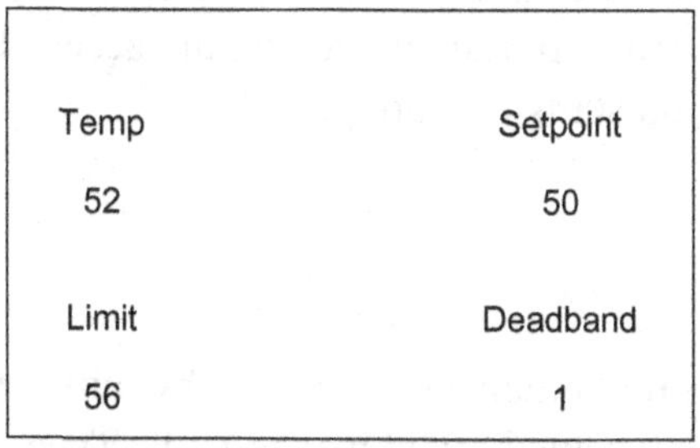

Temp	Setpoint
52	50
Limit	Deadband
56	1

Figure 13.3
Thermostat LCD Screen.

ncurses **Library**

There's actually a higher level solution to this problem that we encountered briefly back in Chapter 6 in the discussion of simulation. The `devices` program makes use of the `ncurses` library to move the cursor around in a window. `ncurses` can also do lots of other things like changing colors and setting fonts.

There are a couple of reasons for not using `ncurses` here. Although it's part of virtually every x86 Linux distribution, it is not by default part of the ARM toolset that we're using. It's a useful tool and I encourage you to check it out, yet nevertheless, understanding the lower level API is still useful.

`ncurses` is part of the GNU library. See Resources for the Web location.

User Space Initialization

We now have a useful thermostat application. The next thing we need to do is arrange for that application to start automatically when the system is powered up. Also, we probably don't need a console shell in the final product.

Let's start by examining the initialization process from the moment power is turned on until the Linux kernel starts the first user space process. There are basically four steps in this process.

Stage 1 Boot Loader

When power is applied to the S3C2440 and the reset pin is released, execution begins at the location specified by the reset vector at address 0. The chip contains 4 KB of SRAM, called the *Steppingstone*, mapped to address 0. When the board is set to boot from NAND flash, the first 4 KB of NAND are copied to the Steppingstone before execution begins. This code does very low-level initialization of things like chip select lines, clocks, and RAM. Basically, the minimum needed to run the processor and access RAM. Then it copies U-boot from Flash to DRAM and starts executing.

U-Boot

The U-boot loader continues the initialization process by initializing the peripherals it needs, typically the serial port, network, and maybe USB. Then, if the auto-boot sequence is not interrupted, it will usually copy the kernel from Flash to RAM and starts executing.

Linux Kernel

If the kernel is a compressed image, it uncompresses itself into the appropriate location in RAM. Then it initializes all of the kernel subsystems such as memory management and drivers for all of the peripheral devices in the system. To some extent, this may duplicate the initialization done by U-boot, but of course the kernel has no idea what U-boot did and so must initialize the devices to suit its own requirements. During the initialization process, the kernel spews out a large number of messages describing what it is doing.

Next, it mounts the root file system. Up to this point the kernel has been running in kernel space. Finally, it starts the init process, which makes the transition to user space.

Init Process

`init` is always the first process that the kernel starts and its job is to get everything up and running properly.

The `init` executable is typically `/sbin/init` although there are several alternative locations that the kernel will search. In the case of our target board, `/sbin/init`, like virtually every other executable, is simply a link back to BusyBox.

`init` gets its instructions from the file `/etc/inittab`. Take a look at `target_fs/etc/inittab`. The comments at the top give a pretty good idea of what's going on. Each entry has four fields separated by colons:

```
<id>:<runlevel>:<action>:process
```

id	tty connected to the process. NULL if the process doesn't use a console.
runlevel	Ignored by BusyBox.
action	One of eight ways init will treat the process.
process	Program to run including arguments if any.

The allowable actions are:

once	Execute the process once.
wait	Execute the process once. Init waits until the process terminates.
askfirst	Ask the user if this process should be run.
sysinit	These processes are executed once before anything else executes.
respawn	Restart the process whenever it terminates.
restart	Like respawn.
shutdown	Execute these processes when the system is shutting down.
ctrlaltdel	Execute this when init receives a SIGINT signal, meaning the operator typed CTRL–ALT–DEL.

Note that the `sysinit` action in this case is to execute the `rcS` script.

Probably, the easiest way to start up our application is from within `inittab`. Add the following entry after the `ttyn` entries:

```
ttySAC0::respawn:/home/src/lcd/netthermo_t
```

We don't need a console terminal, so comment out the existing line that starts with `ttySAC0:`. **Save** `inittab`.

Reset the target and let it boot up Linux. You should see the thermostat start up.

Note, by the way, that another approach to starting the application is to simply replace/ `sbin/init`. Make it a symbolic link that points to the application executable. This may be fine for simple applications, but there is a lot of functionality in init that might be useful in the final product.

In Chapter 14, we'll take a closer look at U-boot and get our "product" ready to "ship."

Resources

From Power Up to Bash Prompt—This HOW-TO from the Linux Documentation Project is a brief description of what happens from the time you turn on power until you log in and get a bash prompt.

www.busybox.net—This is the official BusyBox web site.

A pair of files in the Linux `Documentation/` directory offer additional information on console and framebuffer drivers:

```
serial_console.txt
```

```
framebuffer.txt
```

www.gnu.org/software/ncurses/ncurses.html—Home page of the ncurses project at the Free Software Foundation.

## CHAPTER 14

U-Boot Boot Loader and Getting Ready to Ship

The box said "Requires Windows 95 or better." So, I installed Linux.

U-Boot

As we saw in the previous chapter, a boot loader is a program that does some initialization in preparation for loading and running the OS and its supporting infrastructure. In a sense, the boot loader is like the BIOS in a desktop PC or server. The principal difference is that a boot loader executes once when the system powers up and then goes away. The BIOS hangs around to provide low-level I/O services.

Background

Desktop Linux systems have a boot loader in addition to the BIOS. These days it's usually GRUB. But because the BIOS does all the heavy lifting of initializing the hardware, GRUB itself does little more than load and start the OS. If you're building an embedded system based on conventional x86 PC hardware, GRUB is probably the way to go.

Our target board uses a very popular, very capable, Open Source, cross-platform boot loader to get things started. We briefly saw some of the features of U-Boot back in Chapter 5 when we brought the target board up.

U-Boot began as a PowerPC boot loader named 8xxROM written by Magnus Damm. Wolfgang Denk subsequently moved the project to Source Forge and renamed it PPCBoot. The source code was briefly forked into a product called ARMBoot by Sysgo GmbH. The name was changed to U-Boot when the ARMBoot fork was merged back into the PPCBoot tree. Today, U-Boot supports roughly a dozen architectures and over 400 different boards.

The development of U-Boot is closely tied to Linux with which it shares some header files. Some of the source code originated in the kernel source tree.

U-Boot supports an extensive command set that not only facilitates booting but also manages flash memory, downloads files over the network, and more. Appendix A details the command set. The command set is augmented by environment variables and a scripting language.

Installing and Configuring U-Boot

See the Resources section for the location of the U-Boot source tarball. There are a number of versions to choose from, I happen to have settled on version 1.3.2. Untar the tarball in your home directory. You'll find a new subdirectory named uboot-<version_number>. There's a very extensive README file in the top-level directory. There are additional README files in the doc/ directory, which primarily describe features of specific boards.

U-Boot doesn't support make xconfig or even make menuconfig, at least not yet. Configuration is done by manually editing a header file. cd to include/ configs/ and open mini2440.h. Configuration variables are identified as follows:

- Configuration *options* have names beginning with CONFIG_.
- Configuration *settings*, which are hardware specific, have names beginning with CFG_. These are generally numeric values. The README file cautions that you shouldn't meddle with these "if you don't know what you're doing."
- Included shell commands are identified by names beginning with CONFIG_CMD_.

Like BusyBox, you can select any of the more than 150 shell commands to include in U-Boot. Starting at line 111 is the initial list of included commands starting with a default set. Down at line 299, the NAND partitions are defined followed by the initial setup environment variables.

When you've finished inspecting, and possibly editing, mini2440.h, cd back up to uboot-<version_number>, execute:

```
make mini2440_config
```

The U-Boot Makefile is interesting. Open it with an editor. Starting at line 470, there is a "_config" make target for every header file in the include/configs/ directory. Most of these entries look like the one for the Mini2440 at line 2480. The make target unconfig deletes several files that were created by a previous invocation of the mkconfig script. Next, the mkconfig script is invoked with parameters that describe the board in question.

To build U-Boot then, invoke make by itself just as we've done with the kernel and BusyBox. Note by the way that it's not necessary to specify the cross-tool prefix, arm-linux-. That's taken care of by the configuration process. Of course, if your prefix happens to be different, you'll either have to edit the Makefile to change it or pass the CROSS_COMPILE environment variable as an argument to make like this:

```
make CROSS_COMPILE=my_tool_prefix-
```

The make step generates several output files:

- u-boot—the executable in Executable and Linkable Format (ELF) binary format,
- U-boot.bin—the raw binary image, suitable for writing to flash,
- u-boot.map—map file output by the linker,
- u-boot-nand2k.bin—binary executable padded to a 2k block size,
- u-boot-nand16k.bin—binary executable padded to a 16k block size,
- u-boot.srec—the executable in Motorola S record format.

The make step also builds several tools including mkimage in the tools/ directory.

Testing a New U-Boot

You will want to test your new U-Boot initially in RAM before burning it to flash. You can use a JTAG hardware debugger if you happen to have one. This is useful if you've made substantial changes and really need to test the new U-Boot because it's probably the only way to use breakpoints on boot loader code.

Another approach is similar to what we did in Chapter 5 when we initially configured the board. This approach uses the Supervivi boot loader loaded in NOR flash. Proceed as follows:

1. Copy U-boot.bin to factory_images/.
2. Connect the target board to your workstation with a USB cable.
3. Move the boot select switch to the NOR position and power up (or reset) the board.
4. Select [d] Download and Run from the Supervivi menu.
5. In another shell window, cd to factory_images/ and become a root user.
6. Execute ./boot_usb u-boot.bin.
7. The new U-Boot image should download and start executing.

When you're ready to burn the new boot loader into flash, you can use either the new image or the one currently loaded in flash:

1. Boot into the U-Boot prompt.
2. nand erase u-boot.
3. nand erase env.[1]
4. Move or copy u-boot.bin to /var/lib/tftpboot.
5. tftp 32000000 u-boot.bin.
6. nand write.e 32000000 u-boot $(filesize).[2]
7. Reset. The new U-Boot is running.

---

[1] This step is optional. Doing this forces U-Boot to load the environment with default values. If you're happy with the environment as it is, skip this step.

[2] filesize is an environment variable that is set to the size of the last file downloaded to U-Boot.

"JTAGing" the NOR

OK, let's say your NOR flash gets hosed. What do you do then? Well, the NOR flash can be reprogrammed through the board's JTAG port.

Part of the Mini2440 kit is a device called a JTAG "wiggler" that plugs into a PC's parallel port and connects to the 10-pin flat cable JTAG port on the board. Note that a USB to parallel port converter won't work; it has to be a native parallel port. The directory H-JTAG/ on the kit CD contains a Windows-based application to control the "wiggler." There's also a PDF document there, *JTAGing the NOR*, that explains how to do it.

Creating a Flash File System

We're almost done. We've configured the kernel and the boot loader to our specific needs. We've set up the init process so that it boots directly into the thermostat application. The final step then is to create a YAFFS2 file system image that we can burn into NAND flash.

We saw earlier that the NAND flash is arbitrarily divided into *partitions* of various sizes. Specifically, our flash has four partitions as defined in mini2440.h and listed here for reference:

Name	Size	Offset
u-boot	0 × 00040000	0 × 00000000
env	0 × 00020000	0 × 00040000
kernel	0 × 00500000	0 × 00060000
root fs	0 × 07aa0000	0 × 00560000

The kernel treats these partitions as disk partitions in a device called mtdblock, specifically /dev/mtdblock0 to/dev/mtdblock3. This is what allows us to mount a file system that resides in flash. The kernel also defines a set of character devices, /dev/mtd0 to/dev/mtd3, that allows us to access the flash directly. Execute the following shell command on the target board:

```
head/dev/mtd1
```

This should dump the first part of the U-Boot environment.

The root file system is in the root partition, which translates to mtdblock3. So to mount the root file system from NAND flash, we pass arguments to the kernel on the command line that say mount a YAFFS file system from /dev/mtdblock3.

Run the following commands from the U-Boot prompt:

```
printenv set_bootargs_nand
```

```
run set_booargs_nand
printenv bootargs
```

At the end of Chapter 16, we ended up with a file system that booted directly into our displayable thermostat application. The final step then is to load that file system into NAND flash on the target board so that when the board powers up, the thermostat starts.

We have to create a YAFFS *file system image*, a file in a specific format that contains everything in the file system. There's a utility available from the YAFFS project that does just that. Become root user in a shell window on the workstation. Mount the FriendlyARM DVD, go to `linux/` on the DVD, and copy `mkyaffs2image.tgz` to the root, `/`, directory. `cd/` and untar this file. Two new executables will show up in `/usr/sbin`: `mkyaffs2image` and `mkyaffs2image-128M`. Make sure both files have execute privilege for everyone.

Exit from the root user shell and cd to your home directory. Execute:

```
mkyaffs2image-128M/var/lib/tftpboot/root_qtopia.yaffs root_qtopia
```

This creates the file `root_qtopia.yaffs` in `/var/lib/tftpboot` that is a YAFFS file system image of `root_qtopia/`. There's a potential problem that I've encountered a few times when I've done this. Even though the flash is 128 MB (or more), the entire image file has to fit in the 64 MB of RAM.

If you've been working with the NFS-mounted target file system for a while, you've probably accumulated some junk. Clean all of the directories under `/home/src`. There are 2 MB of pictures in `/opt/Qtopia/pics`. You don't need those. You may have kernel modules under `/lib/modules`. Turns out they're not necessary, the kernel never loads them. You can delete those.

When you have `root_qtopia.yaffs` pared down to under 64 MB, follow these steps, which are similar to what we did with the boot loader:

1. Boot the target board into the U-Boot prompt
2. `tftp 30008000 root_qtopia.yaffs`
3. `nand erase root`
4. `nand write.yaffs 30008000 root $(filesize)`
5. `run set_bootargs_nand`
6. `boot`

The thermostat application should start running.

Finally, you might want to load the kernel you created in Chapter 11. The process is essentially the same as loading the file system. Just copy the kernel `uImage` file to `/var/lib/`

`tftpboot`, substitute that name for "root_qtopia.yaffs" in step 2 above, and substitute "kernel" for "root" in steps 3 and 4.

More Thoughts on Flash Partitions

On the target running Linux, execute `cat/proc/mtd`. You should see the following:

```
mtd0: 00040000 00020000 "supervivi"
mtd1: 00020000 00020000 "param"
mtd2: 00500000 00020000 "Kernel"
mtd3: 07aa0000 00020000 "root"
mtd4: 08000000 00020000 "nand"
```

Compare that with the partition list we got from the `mtdparts` command in U-Boot. That's odd, the names are different. Yes, the kernel has its own view of flash partitions. U-Boot declares the partitions in `include/configs/mini2440.h`. The kernel declares them in arch/arm/mach-2440/mach-mini2440.c. Fortunately, the sizes are the same so the partitions line up. The `00020000` is the partition's erase block size.

There's obviously a potential here for confusion, and even worse. Clearly, it would be preferable to declare the partitions in one place only. Turns out there are a couple of ways to accomplish that. Partitions can be passed on the kernel command line. The format is:

```
mtdparts = <mtddef>[; <mtddef>]
<mtddef>: = <mtd_id>:<partdef>[,<partdef>]
```
where `mtd_id` is a name.
```
<partdef>: = <size>[@offset][<name>][ro][lk]
```
`<size>`: = standard Linux memsize OR "-" to denote all remaining space
```
<name>: = '(' NAME ')'
```

This is essentially the same format as the declaration in `mini2440.h`. In order to do this, you have to turn on a specific configuration variable that is not on by default. That is `CONFIG_MTD_CMDLINE_PARTS`, which is found under `Device Drivers -> Memory Technology Device (MTD) Support -> Command line partition table parsing` in the kernel's `xconfig` menu.

The Flat Device Tree

There's another approach to consistent hardware definition that goes beyond what we've just described and actually defines all of a system's hardware to both the boot loader and the kernel. It's called the *flat device tree* (FDT).

In fact, the problem is more than just the flash partitions. There's a whole host of hardware information that's shared between the boot loader and the kernel:

- RAM—size and location
- Peripherals:
  - Ethernet
  - USB
  - Console

FDT, also known as the Device Tree Blob, device tree binary, or just device tree, provides a mechanism to communicate that information to both the boot loader and the kernel. The FDT is derived from IBM's Open Firmware specification and is a database to represent the low-level hardware details of a board.

A text representation of the hardware is created in a file with the extension `.dts`. A special compiler, `dtc`, converts the text to a binary representation, a "blob," that is loaded into memory along with the kernel, which translates the information to an internal representation used by device drivers to identify hardware resources.

The device tree consists of nodes representing devices or buses. Each node contains properties, `name_value` pairs that give information about the device. The values are arbitrary byte strings, and for some properties, they contain tables or other structured information. We won't go into detail on the syntax here. See the Resources section for references.

The FDT concept evolved in the PowerPC branch of the kernel and that's where it seems to be used the most. In fact, it is now a requirement that all PowerPC platforms pass a device tree to the kernel at boot time. The `.dts` files are typically found in the kernel source tree at `arch/$ARCH/boot/dts`. The powerpc/ branch has 145 `.dts` files while the `arm/` branch only has eight. To get a feel for the syntax, have a look at one of the files in `arch/arm/boot/dts`.

Device tree support has to be built into both the kernel and U-Boot. U-Boot has a couple of configuration variables `CONFIG_OF_LIBFDT` and `CONFIG_OF_BOARD_SETUP` to enable device tree support. The kernel has `CONFIG_DTC` and `CONFIG_OF`. Neither of these is user-settable. They're on if the implementation supports device trees and off otherwise.

The boot process is a little different when device tree support is enabled in U-Boot. You must load both the kernel image and a compiled device tree into RAM. The `bootm` syntax changes to:

```
bootm <kernel_address>-<fdt_address>
```

The <fdt_address> is typically passed in a processor register when control is transferred to the <kernel_address>.

OK, the thermostat application is now burned into flash and we're ready to ship. In the next chapter, we'll turn our attention to source code control.

Resources

www.denx.de—This is the web site for Wolfgang Denk, the principal developer of U-Boot. There's a good user's manual here, a wiki, and of course, the source code at: ftp://ftp.denx.de/pub/u-boot/—Versions of U-Boot going back to 2002 are available here.

www.yaffs.net/—The official site for the YAFFS file system.

www.denx.de/wiki/U-Boot/UBootFdtInfo—Among other things, this page has a good bibliography on flat device trees.

http://ols.fedoraproject.org/OLS/Reprints-2008/likely2-reprint.pdf—*A Symphony of Flavours: Using the Device Tree to Describe Embedded Hardware*, a 2008 Linux Symposium presentation by Grant Likely and Josh Boyer.

## CHAPTER 15

Source Code Control—GIT

And then realize that nothing is perfect.
GIT is just *closer* to perfect than any other SCM out there.

Linus Torvalds

Background

Source Code Control (SCC), also known as Revision Control, refers to the practice of storing the source code files and other artifacts of a project, such as documentation, in a common repository so that multiple developers can work on the project simultaneously without interfering with each other. The SCC software maintains a record of changes and supports multiple versions of a project simultaneously.

There are a number of SCC packages, both Open Source and proprietary, in widespread use. Most of these follow a centralized client/server model where the repositories reside on a central server as illustrated in Figure 15.1. Developers "check out" individual project files, work on them, and subsequently check them back in. Most SCCs allow multiple developers to edit the same file at the same time. The first developer to check in changes to the central repository always succeeds. Many systems provide facilities to automatically *merge* subsequent changes provided they don't conflict.

Some of the better known SCC systems include:

- Concurrent Versioning System (CVS)
- BitKeeper
- Rational Clear Case
- Mercurial
- Perforce
- Subversion
- Visual Source Safe.

Until 2002, the kernel development community didn't use a version control system (VCS). Changes were informally passed around as patches and archive files. In 2002, the community began using BitKeeper, a proprietary distributed VCS. Subsequently, friction between the kernel developers and the BitKeeper people resulted in the former losing their free usage of the package.

DOI: http://dx.doi.org/10.1016/B978-0-12-415996-9.00028-9

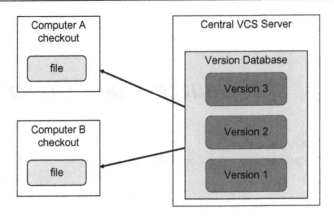

Figure 15.1
Centralized Version Control.

This prompted the Linux community to develop their own tool based on lessons learned while using BitKeeper. The resulting system features a simple design that runs incredibly fast while satisfying these goals:

- Speed
- Simple design
- Fully distributed
- Support for "non-linear" development
- Potentially thousands of parallel branches
- Handles large projects such as the kernel efficiently.

So where did the name "git" come from? Git is British slang for a stupid or unpleasant person. As Linus explained it, "I'm an egotistical bastard, and I name all my projects after myself. First 'Linux', now 'git'."

Introducing Git

Git is a *distributed* VCS where clients don't just check out snapshots of files, they fully mirror the central repository. This is illustrated in Figure 15.2. Each checkout is a full backup. So if the central server dies, it can be restored from any of the clients with little or no loss. Git is installed with the development tools in most modern Linux distributions.

Figure 15.3 shows a typical sequence of git commands. You start by getting a copy of a project repository with the `git clone` command. This creates a complete local copy of the project from the currently active branch of the central repository. The `hello-world` project is 43 versions of the Hello-World program in different languages.

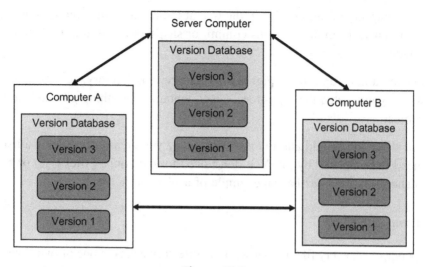

Figure 15.2
Distributed Version Control.

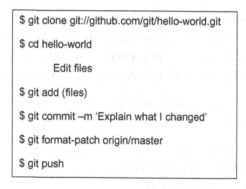

```
$ git clone git://github.com/git/hello-world.git

$ cd hello-world

        Edit files

$ git add (files)

$ git commit –m 'Explain what I changed'

$ git format-patch origin/master

$ git push
```

Figure 15.3
Git Command Sequence.

Then you make changes to the project. `git add` adds the changes you've made to the *index* in preparation for the next *commit*. The index holds a snapshot of the changes that have been added since the last commit. You can invoke `git add` multiple times between commits.

`git commit` stores the index back to the local repository along with a comment describing what has been changed. The –m option means a *message* follows. This is good for relatively short log messages. If you leave off the −m option, git opens `$EDITOR`, allowing you to create a more detailed log message. The significant point is that git insists you provide a comment every time you do a commit.

A *commit* is a snapshot of all the files in the tree at the moment the commit command is executed. Differences from the previous commit or snapshot are stored in a compressed delta form under `.git/objects`.

When you create a new repository (Figure 15.4), a new subdirectory, `.git/`, shows up in the directory where you created the repository. `.git/` contains everything git needs to know about the project.

`.gitignore` is a file you should add to every repository you create. It contains a list of files that don't need to be tracked by git. You don't need to track object files (`*.o`) or backups (`*~`) for instance. `hello-world/` has an example of a `.gitignore` file.

File States and Life Cycle

From git's perspective, every file in your working directory exists in one of two states: tracked or untracked. Tracked files are the ones git knows about and "tracks." Tracked files may be unmodified, modified, or staged. When you clone a project, all the files are initially tracked and unmodified. You determine the state of project files with the `git status` command.

Try this. Add a `README` file to the `hello-world` project and then execute `git status`. The result will look something like Listing 15.1. In this case, there's also a modified file that has been *staged* (by executing git add on it) but hasn't been committed yet. As the status output says, use `git add <file>` to track the file and include it in the next commit.

Execute `git add README`. `README` is now *staged*, that is, ready to be committed, along with any files that have been modified and added since the last commit.

This brings us to the state diagram of Figure 15.5. The primary purpose of this state machine view of git is to make the point that most operations are performed locally, off-line. This has two primary advantages:

1. It's faster.
2. You don't have to be online to work.

```
$ git init              Create a new empty
repository in the current directory

$ git add *      Add all existing files to the index

$ git commit –m 'initial project version'

$ git push      Send to a remote repository
```

Figure 15.4
Creating a New Repository.

```
[doug@ldougs hello-world]$ git status
# On branch master
# Changes to be committed:
#   (use "git reset HEAD <file>..." to unstage)
#
#     modified:   c.c
#
# Untracked files:
#   (use "git add <file>..." to include in what will be committed)
#
#   README
```

Listing 15.1
Git Status.

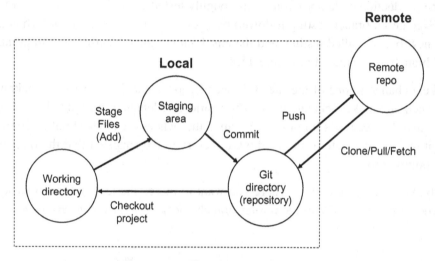

Figure 15.5
Git State Machine View.

We've already seen `git clone` and `git push`. `git fetch` retrieves the updates from the remote repository. `git pull` does a fetch followed by a merge to integrate the updates into the current branch.

Branching and Merging

Perhaps, git's most outstanding feature is its branching model. Other VCSs support branching, but git does so in a particularly elegant way. In fact, some users call the branching model git's "killer feature."

To understand how branching works, we need to step back and take a closer look at how git stores the commits. Figure 15.6 shows the contents of a single commit. It consists of a commit metadata object that contains the author, committer, and commit comment among other things, plus a pointer to a tree object that in turn points to *blobs* representing individual objects that have been changed in this commit. The tree and blobs constitute the *snapshot* that this commit represents.

Figure 15.7 shows the situation after two more commits. Each subsequent commit object points to its *parent*, the commit from which it was created. This is how we can recreate the complete history of the project. There are two other objects in this figure. Master is the name of the current *branch*, the one that you're currently working in. Every project starts with a master branch. HEAD points to the currently active branch. HEAD is a file in .git/.

OK, let's create a new branch. Maybe you want to test out some ideas, but don't want to disturb the production code until you've thoroughly tested them. Try it with the hello-world project. Execute git branch testing followed by git checkout testing. The branch command creates a new branch called testing, and the checkout command causes HEAD to point to the specified branch as illustrated in Figure 15.8.

Now make a change to one of the files followed by git add and git commit. Checkout the master branch and make a change there. The results are depicted in Figure 15.9. The branches have diverged. It's important to realize that whenever to checkout a different branch, git restores the state of the working directory to the snapshot that the branch is currently pointing to.

Eventually, you'll reach the point where you want to merge the testing branch back into the master production branch. With the master branch checked out, execute git merge testing.

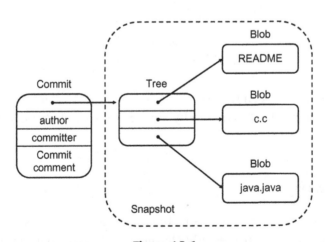

Figure 15.6
Representation of a Commit.

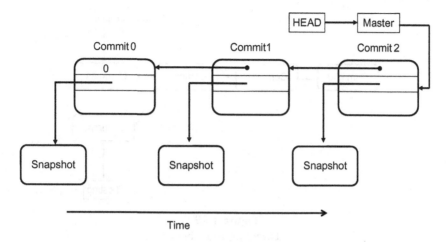

Figure 15.7
Multiple Commits.

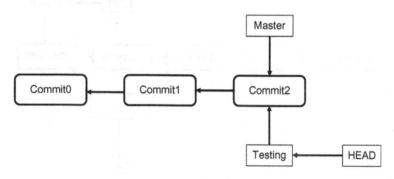

Figure 15.8
Adding a Branch.

Git performs a three-way merge among the two divergent branches and a common ancestor, creating a new commit object. See Figure 15.10.

At this point you probably no longer need the testing branch. You can delete it with `git merge-d testing`.

Configuring Git

When you did your first commit, you may have noticed that git suggested a name and an e-mail address that it pulled from your system information. User name and information are rather important pieces of data because other developers may need to contact you about your commits to the project. So it's important to configure your git package with basic information.

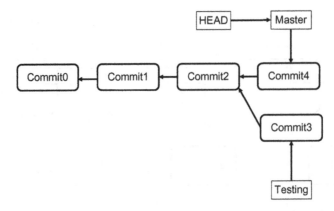

Figure 15.9
Diverging Branches.

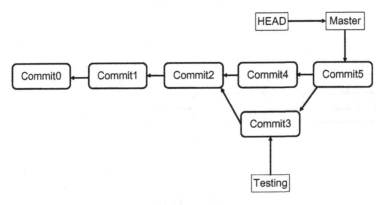

Figure 15.10
Merging Branches.

Git stores configuration information in three places:

1. /etc/gitconfig—System-wide parameters applicable to all users. Use git config
   —system to read or write these values.
2. ~/.gitconfig—Parameters specific to each user. Use git config –global to access these.
3. .git/config—Project-specific parameters found in the repository you're currently
   working with.

Each level overwrites values in the previous level, so values in .git/config take precedence
over those in /etc/gitconfig, for instance. These files are of course plain ASCII text so you
can set the values by manually editing the file and inserting the correct syntax, but it's
generally easier to run the git config command.

To get started, set your user information:

```
git config—global user.name "Homer Simpson"
git config—global user.email homer@thesimpsons.com
```

These commands hint at the basic structure of the configuration files. All configuration variables are specified as `section.variable`. In the configuration file, the section name is enclosed in square brackets. The square brackets may also include a subsection name enclosed in quotes. That syntax becomes `section.subsection.variable`.

To get a better feel for how all of this hangs together, execute `git config—list`, which lists all of the currently defined configuration variables. Then open `hello-world/.git/config` with an editor to see where those values came from.

Git provides a help command, `git help`, which lists all of the commands. If you then execute `git help <command>`, you'll get the man page for `<command>`. Execute `git help config` and you'll eventually come to a list of configuration variables that git recognizes.

Graphical Git

It should perhaps come as no surprise that Eclipse supports git. Fire up Eclipse and change to the Git Repository Exploring perspective. The Project Explorer view has been replaced by Git Repositories view. The menu bar of that view has icons to the following:

- Add a local repository to the view.
- Clone a repository to the view.
- Create a new git repository in the view.

Since we already have a repository for `hello-world`, let's add that to Eclipse. Click the Add icon to bring up the dialog as shown in Figure 15.11. Starting in the `hello-world/` directory, or any parent, click the Search button. Eclipse should find the hello-world repository. Click OK and the repository will be added to the Git Repositories view.

Expand the hello-world entry and various subentries in the Git Repository view to reveal as depicted in Figure 15.12. The context menus for the objects shown here are pretty much what you would expect. If you right-click on Branches you can create a new branch. The currently checked out branch is identified by a check mark. Right-click a different branch name to check it out.

In order to manage hello-world in the context of Eclipse, we need to make it an Eclipse project. Right-click the Working directory entry and select Copy Path to Clipboard. Strictly speaking, this isn't a C/C++ project, so we'll select File > New > Project, expand General and select Project. Click Next and give the project a name. "Hello World" comes to mind.

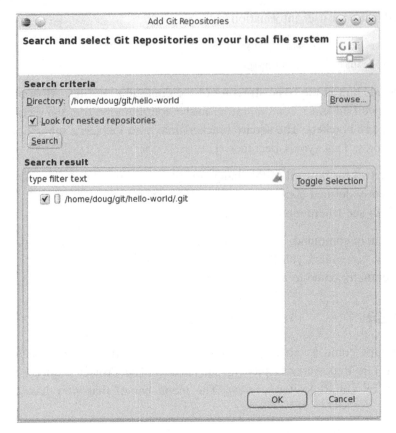

Figure 15.11
Add Repository to Eclipse.

Uncheck Use default location, right-click in the location box, and select Paste. Then click Finish.

Make a change to one or more files. Just double-click the file entries under the Working directory to open them in the editor. Then right-click the changed entries to add them to the staging area. Now right-click the hello-world entry and select Commit. This brings up the dialog as shown in Figure 15.13. The Author and Committer fields are filled out for you and an editor is open to enter the commit message.

Modified files are listed below the Author and the Committer entries. A check mark in status indicates that the file is staged. There's a check mark icon in the tool bar that selects all files. Note that since we created an Eclipse project, there's now a .project file in the working directory. Click Finish and the commit is saved.

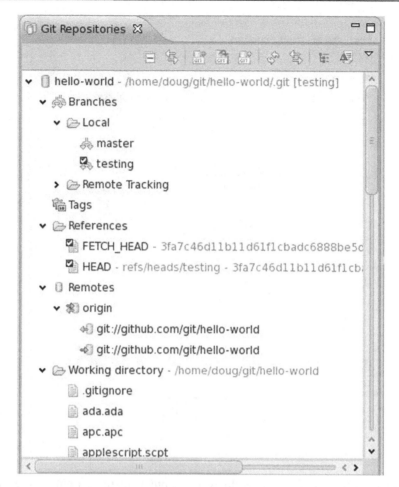

Figure 15.12
Git Repository View.

Creating a New Repository

Ultimately, the point of all this is to bring your own projects under git control. As an example, let's create a new repository out of one of the Eclipse projects we were working with earlier in the book. How about the network project?

To get an existing Eclipse project into a git repository, select the project entry in Project Explorer, right-click and select Team > Share Project. This brings up a dialog where you can select which VCS you want to use. Select git and click next to bring up the dialog as shown in Figure 15.14.

You'll probably want to create a new repository at this point. The question is, where? You can put the repository in the project's folder by checking Use or create Repository

Figure 15.13
Commit Dialog.

in parent folder of project. Or you might want to put it in the parent of the project directories so that you can easily manage all of the projects in one git repository. When you click Create, you get a dialog that asks for a path and a name. The name ends up being a directory under path that must be empty or non-existent. That's where the `.git/` directory is stored. Your project is then moved into the repository directory when you click Finish.

Back in the Project Explorer view, the network entry now has "[my-projects NO-HEAD]." This shows that the project is now managed by the my-projects repository but hasn't had an initial commit yet. If you right-click and select Team, one of the options is Commit, which brings up the Commit dialog that we saw in Figure 15.13. Select all the files, enter a commit message, and click Commit.

You can still do most of your work from the C/C++ perspective. Git doesn't intrude very much on the development process. When you need to do "git like" stuff, such as managing branches, then you need to switch to the Git Repository Exploring perspective.

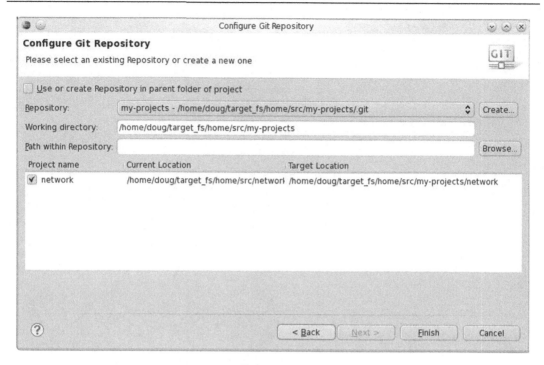

Figure 15.14
Configure Git Repository Dialog.

That wraps up our exploration of SCC using git. In the next and final chapter, we'll look at tools for building complete embedded Linux systems from source.

Resources

http://git-scm.com—The git home page.

http://git-scm.com/book—*Pro Git* by Scott Chacon. This is a very readable and complete introduction to git. The entire book is available in HTML on the git-scm site or you can download it as a PDF.

http://en.wikipedia.org/wiki/Comparison_of_revision_control_software—A very nice comparison of about 30 version control systems including git.

https://github.com—A git hosting service that claims to be hosting over 3 million repositories from almost 2 million people. It's free to Open Source projects. Monthly subscription fees apply to private repositories.

https://na1.salesforce.com/help/doc/en/salesforce_git_developer_cheatsheet.pdf—This is a compact, handy little "cheat sheet" for using git. You can get to it more directly by going to http://git-scm.com/docs.

gitorious.org—Another git hosting service. This one seems to have more of a community feel than github.

CHAPTER 16

Build Tools

The great thing about Object Oriented code is that it can make small, simple problems look like large, complex ones.

An embedded Linux system has several major components, all of which are derived from freely available source code and all of which may require varying levels of customization. These components include:

- Boot loader
- Linux kernel
- Root file system
- Cross-tool chain.

In the course of this book we've treated these as independent elements without paying a lot of attention to where they come from, how they are constructed, or how they interact with each other.

Taken together, these elements constitute an embedded Linux *distribution*. A number of embedded distributions have evolved to address the issues of building and integrating these elements. We'll take a look at some of them in this chapter.

Buildroot

Buildroot grew out of the same project that developed both uClibc and Busybox. uClibc is a smaller footprint version of glibc that was specifically developed for embedded Linux applications. It was originally developed to support uClinux, the variant of Linux for non-Memory Management Unit processors (MMU). Like most contemporary Open Source libraries, uClibc is released under the Library General Public License (LGPL).

Buildroot is a set of Makefiles, scripts, and patches intended to ease the process of building cross-compilation tool chains, root file systems, and Linux kernel images. This is a particularly important issue for embedded developers who are often working with non-x86 processors.

An interesting aspect of Buildroot and other similar build systems such as OpenEmbedded (OE) is that everything is built from sources obtained from the Internet. One consequence of this is that initial builds can take a *long* time, typically several hours. Another feature is

Linux for Embedded and Real-time Applications.
© 2013 Elsevier Inc. All rights reserved.

DOI: http://dx.doi.org/10.1016/B978-0-12-415996-9.00020-4

that Buildroot is specifically designed to be run as a normal user. You don't have to be the root.

Begin by downloading the tarball from buildroot.uclibc.org. At the time of this writing, the most stable release was buildroot-2012.05. Untar it in a suitable location, probably your home directory. Execute `make help` to see what all is available. Note that there is a set of `*_defconfig` target similar to what we saw for the kernel, including `mini2440_defconfig`. So execute `make mini2440_defconfig` followed by `make xconfig`.

Figure 16.1 is a screenshot of the xconfig menu. Broadly speaking, configuration options fall into the following categories:

- Target architecture and variant
- Options for building Buildroot
- Tool chain options
- System Configuration
- Target Package selection
- Boot loader
- Kernel.

The help generally tends to be less helpful than what we saw with kernel configuration. There are a few things you'll probably want to change. Under Toolchain enable Wide CHARacters (WCHAR) support. It seems to be a requirement for a number of other features. Select Linux

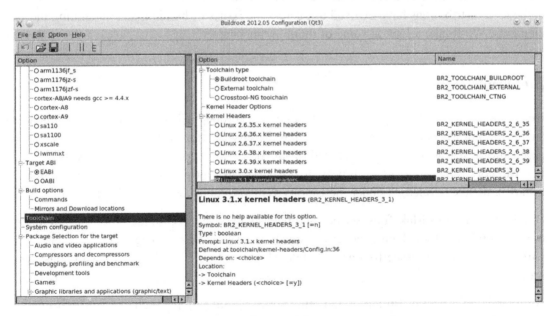

Figure 16.1
BuildRoot xconfig.

3.1.x headers since we're using version 3.1.5. Enable Build gdbserver for the target and Build gdb for the host. Under networking applications you'll probably want to select openssh for the Eclipse Remote System Explorer. Under Filesystem images select ext2 root filesystem as this is what we use for NFS mounting. Exit and save the changes.

Execute `make`. A few hours later you should find the `output/` subdirectory populated like this:

- `build/`—All components except the cross-tool chain are built here.
- `images/`—Final images for the boot loader, kernel, and root file system.
- `host/`—Installation tools compiled for the host needed by BuildRoot including the cross-tool chain.
- `staging/`—A link to `output/host/usr/sysroot`. Its purpose isn't very clear.
- `target/`—An *almost* complete target file system. It's missing the `dev/` directory, which must be created as root user. Use a file system image in `images/` instead.
- `toolchain/`—Build directories for components of the cross-tool chain.

OpenEmbedded

The OE project describes itself as a software framework for creating Linux distributions primarily, but not exclusively, for embedded applications. It grew out of the OpenZaurus project for Sharp Zaurus Personal Digital Assistants (PDAs). The OpenZaurus developers felt they had pushed the limits of Buildroot.

Soon other PDA developers got involved in OE development and over the years many embedded distributions have been based on the OE code base. OE merged with the Yocto project in March 2011 to form the OE-Core project.

Fundamentally, OE comprises two elements: the *build engine*, called BitBake, and the *metadata* that tells BitBake what to do. BitBake is written in Python. To quote the user's manual, "unlike single project tools like make [BitBake] is not based on one makefile or a closed set of interdependent makefiles, but collects and manages an open set of largely independent build descriptions (package recipes) and builds them in proper order."

BitBake's operation can be represented graphically as shown in Figure 16.2. It interprets, or parses if you prefer, recipes and configuration files to determine what needs to be built and how. Then it fetches the necessary source code over the network. The output is a set of packages and file system images.

Metadata can be roughly grouped into four categories, each of which fulfills a specific role:

- Recipes (`*.bb`)—These are the most common form of metadata. A recipe provides instructions for BitBake to build a single package. It describes the package, its dependencies, and any special actions required to build it.

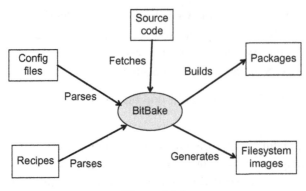

Figure 16.2
BitBake Operation.

- Classes (*.bbclass)—Perform a role similar to that of classes in object-oriented programming languages like C++. They encapsulate common functionality across a large number of recipes.
- Tasks—Used to group packages into a root file system, for example. Tasks are usually relatively simple, consisting of a few lines of package dependencies.
- Configuration (*.conf)—Defines the overall behavior of BitBake.

Getting Started

Bill Sargent's blog (see Resources) has a good HOW-TO on getting started with OE. A number of tools are needed that may or may not be installed on your system. The HOW-TO has a complete list. I chose to install OE directly under my home directory.

We're using a version of OE that's been "tuned" for the Mini2440. Get the repository with this command:

```
git clone git://repo.or.cz/openembedded/mini2440.git
```

When the download is complete you'll have a new directory, mini2440/, containing just over 19,000 files occupying about 400 MB. Have a look at a couple of typical recipe files: recipes/bash/bash_3.0.bb and recipes/helloworld/helloworld_1.0.0.bb. The bash recipe is quite simple. The key word require is equivalent to #include in C. PR is the Package Revision, or version, number. SRC_URI is found in almost every recipe. It tells BitBake how and where to find the files that make up the package.

By contrast, the helloworld recipe lacks an SRC_URI directive and instead defines a do_fetch() method that overrides BitBake's default fetch method. The S and D variables represent respectively the source and destination directories. This recipe also illustrates a couple of additional descriptive variables.

```
mkdir build
mkdir sources
mkdir build/conf
cp mini2440_local_example.conf build/conf/local.conf
```
Edit `conf/local.conf`

   `DL_DIR`—Absolute path to sources directory
   `BBFILES`—Absolute path to recipes
   Note `MACHINE` variable
   Comment `ASSUME_PROVIDED + = gcc3-native`
   Remove line that says "`# REMOVE THIS LINE`"

The last line is a way of making sure you have at least read the entire file.

```
export BBPATH = /path/to/openembedded:/path/to/build/folder
echo 0 >/proc/sys/vm/mmap_min_addr
bitbake—v task-base
```
Have dinner and watch a movie.

This will give you the basic cross compilers, libraries, and tools needed to begin building images or individual packages. Try building the `console-image` recipe:

```
bitbake-v console-image
```

This yields a kernel image, `uImage`, and a root file system, `console-image.jffs2`, that can be burned to flash.

As usual, we only scratched the surface of a very extensive tool. Dig in and learn more if you're so inclined.

Personal Observations

For the most part, I've refrained from inserting my personal biases into this book. Yes, I admitted that I'm a GUI guy and, as a result, I emphasized Eclipse and KDE. But now I'm going to editorialize. I don't like OE. In My Humble Opinion it's unnecessarily complex, poorly documented, and, in my experience, it doesn't work. Granted, your mileage may vary and obviously it works for some people. Nevertheless, I've seen enough dissatisfaction in various forums and mailing lists to reinforce my opinion that OE really isn't ready for prime time.

I recently tried to do an OE build for a Gumstix board. It took over 24 h and generated over 600,000 files and in the end, I wasn't even sure what I had. Oh, and there was a build error that I had to fix by hand. I was actually amazed that the fix worked.

I've had some (but not a lot of) success with BuildRoot. At least BuildRoot is more "approachable" in that it's built from familiar concepts such as scripts and Makefiles.

Android

By now everyone with even the remotest connection to consumer tech knows that Android is Google's implementation of Linux for mobile devices. What you may not know is that Android is now the world's leading smart phone platform with 59% of the market in the first quarter of 2012. One million Android devices are being activated *every day*!

Of course, Android is much more than Linux; it's a complete development framework as illustrated in Figure 16.3. While most Android applications are written in Java, there is no Java Virtual Machine in the platform and Java byte code is not executed. Java classes are compiled into *Dalvik executables* and run on Dalvik, a specialized virtual machine designed specifically for Android and optimized for battery-powered mobile devices with limited memory and CPU.

The most recent release of Android as of this writing is 4.1, Jelly Bean, released in July 2012. Interestingly, since version 1.5, Cupcake, released in April 2009, all releases have been named in alphabetical order for desserts.

There are two distinct levels to Android development: applications and platform development. Application development can be done in Windows, Linux, or Mac OS X (Intel). Platform development is limited to Linux and Mac OS.

Application Development

Begin by downloading the Android Software Development Kit (SDK) for Linux tarball from developer.android.com/sdk/. Untar the 80 some odd megabyte file in a suitable location. This is just the start. Next, you need the Android SDK Platform Tools and an actual platform. Execute `tools/android` to bring up the dialog of Figure 16.4. Select Android SDK Platform tools and Android 4.1 (API 16), then click on Install 7 packages. You'll be asked to accept the license agreements. The simplest thing of course is to click Accept All.

With the downloads complete, the most useful thing to do is go to `docs/about` and open `index.html` with a web browser. This leads you to a very extensive and thorough developer's guide. Eclipse is the preferred development environment for Android, so you'll want to download the Android add-in for Eclipse. After everything is installed and configured, probably the best thing to do is go to Building Your First App.

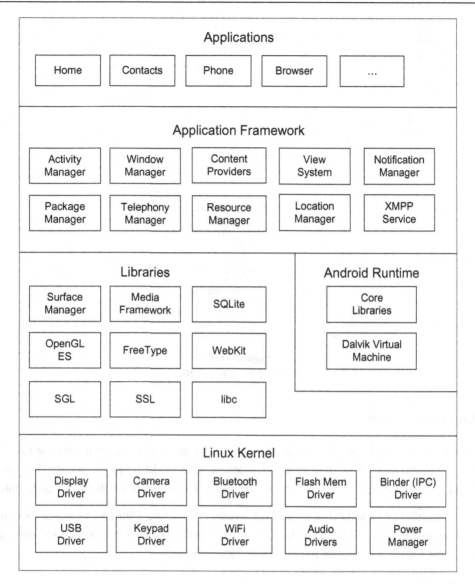

Figure 16.3
Android Framework.

Once you have your first "Hello World" application built, you can run it either on a real Android device connected via USB or on the emulator. Both of these methods can be accessed from Eclipse.

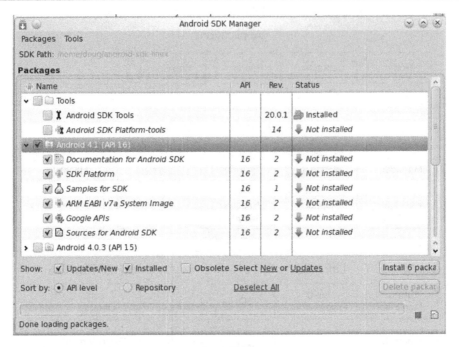

Figure 16.4
Android SDK Manager.

Platform Development

Platform development is another issue altogether. The hardware requirements are rather "steep." Gingerbread (2.3.x) and higher versions require a 64-bit environment. The source download is approximately 8.5 GB, and a single build requires over 30 GB of disk space.

Google routinely tests the Android build on recent versions of Ubuntu but suggests that other Linux distributions should work. Currently, Ubuntu 10.04 is recommended and, again, it must be 64-bit Ubuntu. Other required software includes:

- Python 2.5–2.7
- JDK 6 for building Gingerbread or newer
- JDK 5 for building Froyo or older
- Git 1.7 or newer.

Again, the web site source.android.com has extensive documentation on how to do platform development. Bill Sargent's forum (see Resources) has an extensive HOW-TO on getting started with Android including a rather complete list of required software packages.

Wrap Up

We've covered quite a lot of ground in this tour through the world of embedded Linux. To briefly review, we've covered:

- General overview of Linux in the embedded space including the implications of the GPL,
- Installing Linux and a review of basic features,
- Setting up the development environment and target hardware,
- The Eclipse IDE,
- Application programming in a cross-development environment,
- Configuring and building the Linux kernel and device drivers,
- Various tools and packages such as u-boot, BusyBox, and git version control,
- Embedded Linux distributions and build tools.

I don't pretend that it is in any way complete. We've touched on a number of subjects that require more extensive study if you really want to understand them thoroughly. I hope you'll pursue at least some of the resources that I've listed in the preceding chapters.

Again, remember to sign up for the Yahoo group at groups.yahoo.com/group/embeddedlinuxbook. Happy hacking!

Resources

buildroot.uclibc.org—Home page for Buildroot.
billforums.station51.net/viewtopic.php?f = 3&t = 4—HOW-TO on getting started with OpenEmbedded for the Mini2440.
billforums.station51.net/viewtopic.php?f = 7&t = 13—HOW-TO on getting started with Android on the Mini2440.
git://repo.or.cz/openembedded/mini2440.git—OpenEmbedded repository for the Mini2440.
openembedded.org—Homepage for OpenEmbedded.
source.android.com—Get the Android source code and tools here plus extensive documentation on the development process.

Appendix A: U-Boot Commands

Das U-Boot supports an extensive command set. This section describes the most useful of those commands. Because U-Boot is highly configurable, not all commands are necessarily in any given implementation. The behavior of some commands is also configuration-dependent and may depend on environment-variable values.

All U-Boot commands expect numbers to be entered in hexadecimal. In the following descriptions, arguments are bracketed by < >. Optional arguments are further bracketed by [ ]. Command names can be generally abbreviated to the shortest unique string, shown in parentheses.

Information Commands

bdinfo (bdi)—Displays information about the target board such as memory sizes and locations, clock frequencies, and MAC address. This information is available to the Linux kernel.

coninfo (conin)—Displays information about the available console I/O devices. It shows the device name, flags, and the current usage.

flinfo [<bank #>] (fli)—Displays information about the available flash memory. Without an argument, flinfo lists all flash memory banks. With a numeric argument, n, it lists just memory bank n.

iminfo <start addr> (imi)—Displays header information for images such as Linux kernels or ramdisks. It lists the image name, type, and size and verifies the CRC32 checksum. iminfo takes an argument that is the starting address of the image. The behavior of ininfo is influenced by the verify environment variable.

help [<command name>]—Self-explanatory. Without an argument it prints a short description of all commands. With a command name argument it provides more detailed help.

Memory Commands

By default, the memory commands operate on 32-bit integers. They can be made to operate on 16-bit words or 8-bit types by appending ".w" (for 16 bits) or ".b" (for 8 bits) to the command name. There is also a ".l" suffix to explicitly specify 32 bits. Example: cmp.w 1000 2000 20 compares 32 (20 hex) words at addresses 0 × 1000 and 0 × 2000.

base [<offset>] (ba)—Gets or sets a base address to be used as an offset for all other memory commands. With no argument it displays the current base, default is 0. A numeric argument becomes the new base. This is useful for repeated accesses to a specific section of memory.

crc32 <start addr> <length> [<store addr>] (crc)—Computes and displays a 32-bit checksum over <length> bytes starting at <start addr>. The optional third argument specifies a location to store the checksum.

cmp <addr1> <addr2> <count>—Compares two regions of memory. <count> is the number of *data items* independent of the specified size, byte, word or long.

cp <source> <dest> <count>—Copies a range of memory to another location. <count> is the number of items copied.

md <addr> [<count>]—Memory display. Displays a range of memory in both hex and ASCII. The optional <count> argument defaults to 64 items. md remembers the most recent <addr> and <count> so that if it is entered without an address argument, it continues from where it left off.

mm <start addr>—Modifies memory with auto-increment. Displays the address and current contents of an item and prompts for user input. A valid hexadecimal value replaces the current value and the command displays the next item. This continues as long as you enter valid hex values. To terminate the command, enter a non-hex value.

mtest <start addr> <end addr> [<pattern>]—Simple read/write RAM test. This test modifies RAM and may crash the system if it touches areas used by U-Boot such as the stack or heap.

mw <addr> <value> [<count>]—Writes <value> to a range of memory starting at <addr>. The optional <count> defaults to 1.

nm <addr>—Modifies memory at a constant address (no auto-increment). Interactively writes valid hex data to the same address. This is useful for accessing I/O device registers. Terminate the command with a non-hex input.

loop <addr> <count>—Reads memory in a tight loop. This can be useful for scoping. This command **never terminates!** The only way out is to reset the board.

NOR Flash Memory Commands

U-Boot organizes NOR flash memory into *sectors* and *banks*. A bank is defined as an area of memory consisting of one or more memory chips connected to the same *chip select* signal. A bank is subdivided into sectors where a sector is the smallest unit that can be written or erased in a single operation. Bank numbering begins at 1 while sectors are numbered starting with 0.

`cp <source> <dest> <count>`—The `cp` command is "flash aware." If `<dest>` is in flash it invokes the appropriate programming algorithm. Note that the copy may fail if the target area has not been erased or if it includes any protected sectors.

`erase <start> <end> (era)`—Erases flash from `<start>` to `<end>`. `<start>` must be the first address of a sector and `<end>` must be the last address of a subsequent sector. Otherwise the command does not execute. A warning is displayed if the range includes any protected sectors. The range to be erased can also be expressed in banks and sectors.

`erase <bank #>:<start sector>[-<end sector>]`—Erases from `<start sector>` to `<end sector>` in flash bank `<bank #>`. If `<end sector>` is not present, only `<start sector>` is erased.

`erase all`—Erases all flash memory except protected sectors.

`protect on | off <start> <end>`—Sets write protection on or off for the specified range. `<start>` must be the first address of a sector and `<end>` must be the last address of a subsequent sector. Like the erase command, the protect range can be expressed in banks and sectors.

`protect on | off <bank #>:<start sector>[-<end sector>]`—Sets write protection on or off from `<start sector>` to `<end sector>` in flash bank `<bank #>`. If `<end sector>` is not present, only `<start sector>` is affected.

`protect bank <bank #>`—Sets write protection on or off for all sectors in the specified bank.

`protect all`—Sets write protection on or off for all flash in the system.

Note that the protection mechanism is software-only that protects against writing by U-Boot. Additional hardware protection depends on the capabilities of the flash chips and the flash device driver.

NAND Flash Memory Commands

NAND flash is organized into arbitrary sized named *partitions*. All NAND commands are prefixed by the key word `nand`.

`nand info`—Shows available NAND devices.

`mtdparts`—Lists NAND partitions.

`nand erase [clean] [<offset> <size>]`—Erases `<size>` bytes starting at `<offset>`. If `<offset>` is not specified, erases entire chip.

nand scrub—*Really* erases the entire chip, including bad blocks. Considered "unsafe."

nand createbbt—Creates bad block table.

nand bad—Lists bad blocks.

nand read[.jffs2|.yaffs]<address><offset>|<partition><size>—Reads <size> bytes from flash either from numerical <offset> or <partition> name into RAM at <address>. The .jffs2 and .yaffs modifiers identify file system images.

nand write[.e|.jffs2|.yaffs]<address><offset>|<partition><size>—Writes <size> bytes from <address> in RAM to flash either at numerical <offset> or <partition> name. The .e modifier means write the error correction bytes. The .jffs2 and .yaffs modifiers identify file system images.

Execution Control Commands

autoscr <addr>—Executes a "script" of U-Boot commands. The script is written to a text file that is then converted to a U-Boot image with the mkimage utility. <addr> is the start of the image header.

bootm <addr> [<param> ...]—Boots an image such as an OS from memory. The image header, starting at <addr>, contains the necessary information about the OS type, compression, if any, load and entry point addresses. <param> is one or more optional parameters passed to the OS. The OS image is copied into RAM and uncompressed if necessary. Then control is transferred to the entry point address. For Linux, one optional parameter is recognized, the address of an initrd RAM disk image.

Note incidentally that images can be booted from RAM, having been downloaded, for example, with TFTP. In this case, be careful that the compressed image does not overlap the memory area used by the uncompressed image.

go <addr> [<param> ...]—Transfers control to a "stand-alone" application starting at <addr> and passing the optional parameters. These are programs that don't require the complex environment of an OS.

Download Commands

Three commands are available to boot images over the network using TFTP. Two of them also obtain an IP address before executing the file download.

bootp <load_addr> <filename>—Obtain an IP address using the bootp protocol, then download <filename> to <load_addr>.

`rarpboot <load_addr> <filename> (rarp)`—Obtain an IP address using the RARP protocol, then download `<filename>` to `<load_addr>`.

`tftpboot <load_addr> <filename> (tftp)`—Just download the file. Assumes client already has an IP address, either statically assigned or obtained through DHCP.

`dhcp`—Get an IP address using DHCP.

It's also possible to download files using a serial line. The recommended terminal emulator for serial image download is kermit as some users have reported problems using minicom for image download.

`loadb <offset>`—Accept a binary image download to address `<offset>` over the serial port. Start this command in U-Boot, then initiate the transmission on the host side.

`loads <offset>`—Accept an S-record file download to address `<offset>` over the serial port.

Environment Variable Commands

`printenv [<name> ...]`—Prints the value of one or more environment variables. With no argument, `printenv` lists all environment variables. Otherwise, it lists the value(s) of the specified variable(s).

`setenv <name> [<value>]`—With one argument, `setenv` removes the specified variable from U-Boot's environment and reclaims the storage. With two arguments it sets variable `<name>` to `<value>`. These changes take place in RAM only. **Warning:** use a space between <name> and <value>, not " = ". The latter will be interpreted literally with rather strange results.

Standard shell quoting rules apply when a value contains characters with special meaning to the command line parser such as "$" for variable substitution and ";" for command separation. These characters are "escaped" with a backslash, "\". Example:

```
setenv netboot tftp 21000000 uImage\; bootm
```

`saveenv`–Writes the environment to persistent storage.

`run <name> [...]`—Treats the value of environment variable `<name>` as one or more U-Boot commands and executes them. If more than one variable is specified, they are executed in order.

`bootd (boot)`—A synonym for run bootcmd to execute the default boot command.

Environment Variables

The U-Boot environment is kept in persistent storage and copied to RAM when U-Boot starts. It stores environment variables used to configure the system. The environment is protected by a CRC32 checksum. This section lists some of the environment variables that U-Boot recognizes.

The variables shown here serve specific purposes in the context of U-Boot and, for the most part, will not be used explicitly. When needed, U-Boot environment variables are used explicitly in commands much the same way that they are in shell scripts and makefiles, by enclosing them in $().

autoload—If set to "no," or any string beginning with "n", the rarp and bootp commands will only get configuration information and not try to download an image using TFTP.

autostart—If set to "yes," an image loaded using the rarpb, bootp, or tftp commands will be automatically started by internally calling the bootm command.

baudrate—A decimal number that specifies the bit rate for the console serial port. Only a predefined list of baud-rate settings is available. Following the setenv baudrate <n> command, U-Boot expects to receive a newline character at the new rate before actually committing the new rate. If this fails, the board must be reset and reverts to the old baud rate.

bootargs—The value of this variable is passed to the Linux kernel as boot arguments, i.e., the "command line."

bootcmd—Defines a command string that is automatically executed when the initial countdown is *not* interrupted, but only if the variable bootdelay is also defined with a non-negative value.

bootdelay—Wait this many seconds before executing the contents of the bootcmd variable. The delay can be interrupted by pressing any key before the bootcmd sequence starts. **Warning:** setting bootdelay to 0 executes bootcmd immediately and effectively disables any interaction with U-boot. On the other hand, setting this variable to −1 disables auto boot.

bootfile—Name of the default image to be loaded by the tftpboot command.

ethaddr—MAC address for the first or only Ethernet interface on the board, known to Linux as eth0. A MAC address is 48 bits represented as six pairs of hex digits separated by dots.

eth1addr, eth2addr—MAC addresses for the second and third Ethernet interfaces when present.

ipaddr—Sets an IP address on the target board for TFTP downloading.

loadaddr—Default buffer address in RAM for commands like tftpboot, loads, and bootm. Note: it appears, but does not seem to be documented, that there is a default load address, 0x21000000, built into the code.

serverip—IP address of the TFTP server used by the tftpboot command.

serial#—A string containing hardware identification such as type and serial number. This variable can only be set once, often during manufacturing of the board. U-Boot refuses to delete or overwrite this variable once it has been set.

verify—If set to "n" or "no", disables the checksum calculation over the complete image in the bootm command to trade speed for safety in the boot process. The header checksum is still verified.

The following environment variables can be automatically updated by the network boot commands (bootp, dhcp, or tftp) depending on the information provided by your boot server:

bootfile—see above,

dnsip—IP address of your Domain Name Server

gatewayip—IP address of the Gateway (Router),

hostname—target hostname.

ipaddr—see above,

netmask—subnet mask,

rootpath—path to the root filesystem on the NFS server,

serverip—see above,

filesize—size in bytes, as a hex string, of the file downloaded using the last bootp, dhcp, or tftp command.

Appendix B: Why Software Should Not Have Owners

Digital information technology contributes to the world by making it easier to copy and modify information. Computers promise to make this easier for all of us.

Not everyone wants it to be easier. The system of copyright gives software programs "owners", most of whom aim to withhold software's potential benefit from the rest of the public. They would like to be the only ones who can copy and modify the software that we use.

The copyright system grew up with printing—a technology for mass production copying. Copyright fit in well with this technology because it restricted only the mass producers of copies. It did not take freedom away from readers of books. An ordinary reader, who did not own a printing press, could copy books only with pen and ink, and few readers were sued for that.

Digital technology is more flexible than the printing press: when information has digital form, you can easily copy it to share it with others. This very flexibility makes a bad fit with a system like copyright. That's the reason for the increasingly nasty and draconian measures now used to enforce software copyright. Consider these four practices of the Software Publishers Association (SPA):

- Massive propaganda saying it is wrong to disobey the owners to help your friend.
- Solicitation for stool pigeons to inform on their coworkers and colleagues.
- Raids (with police help) on offices and schools, in which people are told they must prove they are innocent of illegal copying.
- Prosecution (by the US government, at the SPA's request) of people such as MIT's David LaMacchia, not for copying software (he is not accused of copying any), but merely for leaving copying facilities unguarded and failing to censor their use.

All four practices resemble those used in the former Soviet Union, where every copying machine had a guard to prevent forbidden copying, and where individuals had to copy information secretly and pass it from hand to hand as "samizdat". There is of course a difference: the motive for information control in the Soviet Union was political; in the US the motive is profit. But it is the actions that affect us, not the motive. Any attempt to block the sharing of information, no matter why, leads to the same methods and the same harshness.

Owners make several kinds of arguments for giving them the power to control how we use information:

- Name calling.

    Owners use smear words such as "piracy" and "theft", as well as expert terminology such as "intellectual property" and "damage", to suggest a certain line of thinking to the public—a simplistic analogy between programs and physical objects.

    Our ideas and intuitions about property for material objects are about whether it is right to *take an object away* from someone else. They don't directly apply to *making a copy* of something. But the owners ask us to apply them anyway.

- Exaggeration.

    Owners say that they suffer "harm" or "economic loss" when users copy programs themselves. But the copying has no direct effect on the owner, and it harms no one. The owner can lose only if the person who made the copy would otherwise have paid for one from the owner.

    A little thought shows that most such people would not have bought copies. Yet the owners compute their "losses" as if each and every one would have bought a copy. That is exaggeration—to put it kindly.

- The law.

    Owners often describe the current state of the law, and the harsh penalties they can threaten us with. Implicit in this approach is the suggestion that today's law reflects an unquestionable view of morality—yet at the same time, we are urged to regard these penalties as facts of nature that can't be blamed on anyone.

    This line of persuasion isn't designed to stand up to critical thinking; it's intended to reinforce a habitual mental pathway.

    It's elementary that laws don't decide right and wrong. Every American should know that, forty years ago, it was against the law in many states for a black person to sit in the front of a bus; but only racists would say sitting there was wrong.

- Natural rights.

    Authors often claim a special connection with programs they have written, and go on to assert that, as a result, their desires and interests concerning the program simply outweigh those of anyone else—or even those of the whole rest of the world. (Typically companies, not authors, hold the copyrights on software, but we are expected to ignore this discrepancy.)

    To those who propose this as an ethical axiom—the author is more important than you—I can only say that I, a notable software author myself, call it bunk.

    But people in general are only likely to feel any sympathy with the natural rights claims for two reasons.

One reason is an overstretched analogy with material objects. When I cook spaghetti, I do object if someone else eats it, because then I cannot eat it. His action hurts me exactly as much as it benefits him; only one of us can eat the spaghetti, so the question is, which? The smallest distinction between us is enough to tip the ethical balance.

But whether you run or change a program I wrote affects you directly and me only indirectly. Whether you give a copy to your friend affects you and your friend much more than it affects me. I shouldn't have the power to tell you not to do these things. No one should.

The second reason is that people have been told that natural rights for authors is the accepted and unquestioned tradition of our society.

As a matter of history, the opposite is true. The idea of natural rights of authors was proposed and decisively rejected when the US Constitution was drawn up. That's why the Constitution only *permits* a system of copyright and does not *require* one; that's why it says that copyright must be temporary. It also states that the purpose of copyright is to promote progress—not to reward authors. Copyright does reward authors somewhat, and publishers more, but that is intended as a means of modifying their behavior.

The real established tradition of our society is that copyright cuts into the natural rights of the public—and that this can only be justified for the public's sake.

• Economics.

The final argument made for having owners of software is that this leads to production of more software.

Unlike the others, this argument at least takes a legitimate approach to the subject. It is based on a valid goal—satisfying the users of software. And it is empirically clear that people will produce more of something if they are well paid for doing so.

But the economic argument has a flaw: it is based on the assumption that the difference is only a matter of how much money we have to pay. It assumes that "production of software" is what we want, whether the software has owners or not. People readily accept this assumption because it accords with our experiences with material objects. Consider a sandwich, for instance. You might well be able to get an equivalent sandwich either free or for a price. If so, the amount you pay is the only difference. Whether or not you have to buy it, the sandwich has the same taste, the same nutritional value, and in either case you can only eat it once. Whether you get the sandwich from an owner or not cannot directly affect anything but the amount of money you have afterwards.

This is true for any kind of material object—whether or not it has an owner does not directly affect what it *is*, or what you can do with it if you acquire it.

But if a program has an owner, this very much affects what it is, and what you can do with a copy if you buy one. The difference is not just a matter of money. The system of owners of software encourages software owners to produce something—but not what society really needs. And it causes intangible ethical pollution that affects us all.

What does society need? It needs information that is truly available to its citizens—for example, programs that people can read, fix, adapt, and improve, not just operate. But what software owners typically deliver is a black box that we can't study or change.

Society also needs freedom. When a program has an owner, the users lose freedom to control part of their own lives.

And above all society needs to encourage the spirit of voluntary cooperation in its citizens. When software owners tell us that helping our neighbors in a natural way is "piracy", they pollute our society's civic spirit.

This is why we say that free software is a matter of freedom, not price.

The economic argument for owners is erroneous, but the economic issue is real. Some people write useful software for the pleasure of writing it or for admiration and love; but if we want more software than those people write, we need to raise funds.

For ten years now, free software developers have tried various methods of finding funds, with some success. There's no need to make anyone rich; the median US family income, around $35k, proves to be enough incentive for many jobs that are less satisfying than programming.

For years, until a fellowship made it unnecessary, I made a living from custom enhancements of the free software I had written. Each enhancement was added to the standard released version and thus eventually became available to the general public. Clients paid me so that I would work on the enhancements they wanted, rather than on the features I would otherwise have considered highest priority.

The Free Software Foundation (FSF), a tax-exempt charity for free software development, raises funds by selling GNU CD-ROMs, T-shirts, manuals, and deluxe distributions (all of which users are free to copy and change), as well as from donations. It now has a staff of five programmers, plus three employees who handle mail orders.

Some free software developers make money by selling support services. Cygnus Support, with around 50 employees [when this article was written], estimates that about 15 per cent of its staff activity is free software development—a respectable percentage for a software company.

Companies including Intel, Motorola, Texas Instruments and Analog Devices have combined to fund the continued development of the free GNU compiler for the language C. Meanwhile, the GNU compiler for the Ada language is being funded by the US Air Force, which believes this is the most cost-effective way to get a high quality compiler. [Air Force funding ended some time ago; the GNU Ada Compiler is now in service, and its maintenance is funded commercially.]

All these examples are small; the free software movement is still small, and still young. But the example of listener-supported radio in this country [the US] shows it's possible to support a large activity without forcing each user to pay.

As a computer user today, you may find yourself using a proprietary program. If your friend asks to make a copy, it would be wrong to refuse. Cooperation is more important than copyright. But underground, closet cooperation does not make for a good society. A person should aspire to live an upright life openly with pride, and this means saying "No" to proprietary software.

You deserve to be able to cooperate openly and freely with other people who use software. You deserve to be able to learn how the software works, and to teach your students with it. You deserve to be able to hire your favorite programmer to fix it when it breaks.

You deserve free software.

Updated: $Date: 2001/09/15 20:14:02 $ $Author: fsl $

Index

Note: Page numbers followed by "*f*" and "*t*" refer to figures and tables, respectively.

Printed in the United States
By Bookmasters